Springer-Verlag France S.A.R.L

Springer-Verlag France

26, rue des Carmes, 75005 Paris, France

Jean-Claude Rambaud   J. Thomas LaMont   (Hrsg.)

Ökosystem Darm *Special*

Updates on *Clostridium difficile*

Mit 33 Abbildungen und 21 Tabellen

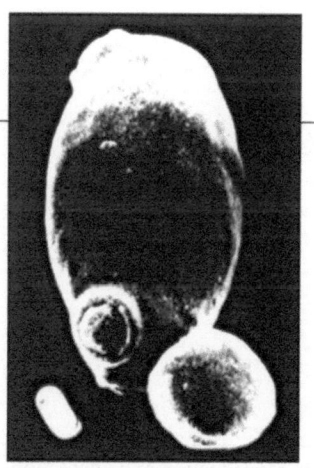

Updates on
Clostridium difficile

Paris, Mai 1995

Springer

Professeur J.-C. Rambaud
Service d'Hépato-Gastro-Entérologie
Hôpital Saint-Lazare
107, rue du faubourg Saint-Denis
75475 Paris Cedex 10
FRANCE

Professor J.T. LaMont
Division of Gastroenterology
Beth Israel Hospital
330 Brookline Avenue
Boston, MA 02215
USA

© Springer-Verlag France 1996
Originally published by Springer-Verlag France, Paris in 1996

ISBN 978-2-287-59639-1   ISBN 978-2-8178-0924-3 (eBook)
DOI 10.1007/978-2-8178-0924-3
Hong Kong Londres Milan Santa Clara
Printed on acid-free paper

Proceedings of the Symposium
« Updates on *Clostridium difficile* »

Comptes rendus du Symposium
« Acquisitions récentes sur *Clostridium difficile* »

Zusammenfassungen des Symposiums
« Updates on *Clostridium difficile* »

Paris – Hôtel Méridien Étoile
Friday, 5th of May 1995

Chairmen
J.-C. Rambaud and J.T. LaMont

Moderators
P. Rampal (Nice), M. Cerf (Paris),
G. Corthier (Jouy-en-Josas), T.D. Wilkins (Blacksburg)
E. Bergogne-Bérézin (Paris)

Sponsored by
La Société Nationale Française de Gastro-Entérologie

Table of contents
Table des matières
Inhaltsverzeichnis

List of contributors
Liste des auteurs
Autorenliste

Frédéric Barbut
   Service de Bactériologie-Virologie, Hôpital Saint-Antoine,184 rue du Faubourg Saint-Antoine, 75571 Paris Cedex 12, France

Richard G. Bennett
   Johns Hopkins Geriatrics Center, 5505 Johns Hopkins Bayview Circle, Baltimore, Maryland 21224, USA

Jean-Paul Buts
   Unité de Gastro-Entérologie Pédiatrique, Université Catholique de Louvain, Cliniques Universitaires St. Luc - a.s.b.l., Avenue Hippocrate 10/1301, B-1200 Bruxelles, Belgique

Michel Delmée
   Unité de Microbiologie, Université Catholique de Louvain, Faculté de Médecine, Avenue Hippocrate 54, U.C.L. 54.90, B-1200 Bruxelles, Belgique

Robert Fekety
   Department of Internal Medecine, Division of Infectious Diseases, 3116 Taubman Center, University of Michigan Hospital, Ann Arbor, Michigan 48109, USA

J. Thomas LaMont
   Division of Gastroenterology, Beth Israel Hospital, 330 Brookline Avenue, Boston, Massachussetts 02215, USA

Lynne V. McFarland
   Department of Medicinal Chemistry, School of Pharmacy, University of Washington, Seattle, Washington 98195, USA

Philippe Marteau
   Service d'Hépato-Gastro-Entérologie, Hôpital Saint-Lazare, 107 bis rue du Faubourg Saint-Denis, 75475 Paris Cedex 10, France

Jean-Claude Petit
   Service de Bactériologie-Virologie, Hôpital Saint-Antoine,184 rue du Faubourg Saint-Antoine, 75571 Paris Cedex 12, France

Charalabos Pothoulakis
Division of Gastroenterology, Beth Israel Hospital, 330 Brookline Avenue, Boston, Massachusetts 02215, USA

Jean-Claude Rambaud
Service d'Hépato-Gastro-Entérologie, Hôpital Saint-Lazare, 107 bis rue du Faubourg Saint-Denis, 75475 Paris Cedex 10, France

Christina M. Surawicz
Harborview Medical Center, Division of Gastroenterology, 325 Ninth Avenue, Seattle, Washington 98 104, USA

Soad Tabaqchali
Department of Medical Microbiology, St. Bartholomew's Hospital Medical College, West Smithfield, London EC1A 7BE, Great-Britain

Michel Warny
Unité de Microbiologie, Université Catholique de Louvain, Faculté de Médecine, Avenue Hippocrate 54, U.C.L. 54.90, B-1200 Bruxelles, Belgique

Mark Wilks
Department of Medical Microbiology, St. Bartholomew's Hospital Medical College, West Smithfield, London EC1A 7BE, Great-Britain

Introduction

Jean-Claude Rambaud

The place occupied today in basic and clinical research by intestinal disease related to *Clostridium difficile* is such that it is hard to remember that this range of disorders was completely identified only in 1977-1978, even though pieces of the puzzle had been identified much earlier. A brief historical review of the discovery of the enteropathogenicity of *C. difficile* in man might thus be useful.

The bacterium was described in 1935 in the stools of infants, using the name *Bacillus difficilis* [7]. Until 1977, the microorganism, renamed *C. difficile*, considered to be of endogenous origin, was isolated only in rare cases of abscess or infection, most often unrelated to the digestive tract. Its role in genito-urinary infections [6] was not confirmed. However, the frequency of infant healthy carriers was recognized from the outset [7, 13].

Pseudo-membranous colitis (PMC) was described in 1883 following a gastroenterostomy. Many cases of this condition were published subsequently before the antibiotic era, describing various risk factors [4]. However the disease began to flourish only with the increasingly wide use of antibiotics. Antibiotic-associated PMC was first described as an enterocolitis, though with little pathological evidence. It was principally related to the use of chloramphenicol and tetracyclines and attributed to proliferation of *Staphylococcus aureus* [11], a concept strengthened by the spectacular therapeutic action of vancomycin. The reason for which PMC only rarely complicates treatment with these two antibiotics today, but on the contrary lincosamines, broad spectrum penicillins and cephalosporins, is probably due to changes in the susceptibility to antibiotics of *C. difficile* and above all of the barrier flora. The occurrence in 1974 of an epidemic of PMC confirmed by sigmoidoscopy and unrelated to *S. aureus* [14] gave rise to renewed interest in the etiology of the disorder. It is interesting to note that the causal role of *C. difficile* and its toxins in these cases was established retrospectively in stool samples kept since 1973 [1]. Evidence was reported in 1977 of a cytopathic effect of acellular supernatants of stools of patients with PMC, this effect being attributed to a possible virus [10].

The lethal effect of several antibiotics, including initially penicillin, on some rodents was noted as early as 1943 [8]. Such acute toxicity in the animal would probably have halted the development of penicillin today! The only lesion constantly found at autopsy in the guinea pig and hamster was a fluid-filled dilated cecum and the now well-known enteric lesions. Independently, an epidemic of fatal enterocolitis in the chinchilla, seen in the early 1950s [15], was linked to the presence of chlortetracycline in food but was very probably erroneously considered to be due to *S. aureus*. In experimental studies of antibiotic-related enterocolitis, the hamster treated with lincomycin (then clindamycin) proved to be a particularly useful model [12], later used by Bartlett et al [3], due to the susceptibility of this animal to lincosamines. Evidence was found in 1974 of a cytopathogenic effect of the acellular supernatant of the intestinal contents of penicillin-treated guinea pigs and hamsters, an effect concerning many cell types and attributed to a probable virus [5].

It was finally in the model of the clindamycin-treated hamster that the discovery of the protective effect of oral vancomycin focused research on a Gram + bacterium and led in 1977 to the isolation of a strain of *C. difficile* [3]. One year later, this same bacterium was isolated in feces from a case of PMC [2]. This bacterium was responsible, via toxin production, for the cytopathogenic effect of feces seen in 1974 in the animal and in 1977 in man.

What can be learned from this detective story and its happy ending, other than that research benefits greatly from an interdisciplinary approach? It is likely that if the attention of clinicians and pathologists had been drawn sooner to antibiotic-associated enterocolitis in rodents, the model used successfully by Bartlett would have led to the earlier discovery of the toxinogenic bacterium responsible for PMC. It is also possible that the role of «*Bacillus difficilis*» and of its « neurotoxin » - biological activity actually borne by toxins A and B - might have been considered sooner, if the presence of this pathogenic bacterium had been routinely sought. However, at any event, the cause of PMC would never have been discovered without the development of antibiotic treatment, responsible not only for multiplication of the human disease but also for the creation of useful animal models.

Less than 20 years after the achievements of this pioneer work, there have been exponential advances in all domains: clinical, epidemiological, microbiological, pathophysiological and therapeutic. Extensively concentrated on microbiological and pathophysiological findings and their therapeutic consequences, the first symposium on *C. difficile* and Intestinal Diseases, just 5 years ago, can be seen as the point of departure of a concern focused much more on the preoccupations of clinicians, though without leaving more basic aspects in the shadows. Thus the clinico-pathological pattern of intestinal disease related to *C. difficile* has been defined and broadened, from ordinary diarrhea to toxic megacolon, its special features at the two extremes of life have more clearly emerged, microbiological diagnostic methods, which were painfully inadequate, have become more widely available, enabling incidence studies and other epidemiological surveys and, finally, appropriate treatment, both antibiotic and ecological, has been defined.

This meeting would nevertheless have lost much of its value if the latest developments in basic research on *C. difficile*, the essential background to future clinical and therapeutic advances, were ignored. The study of epidemiological markers, the number of which requires selection and standardization, understanding of the mechanism of action of toxins A and B, and particularly of their mode of attachment and their intracellular target, investigation of immunity regarding toxins, the extent of which must be envisaged in view of the propensity to recurrences, and elucidation of the mechanism of action of *Saccharomyces boulardii* in recurrent PMC, are far from complete. Perhaps consideration should have been given to the bacterial barrier against *C. difficile* and the reasons for its failure to redevelop after antibiotic treatment in some patients. Further attention should also certainly be paid to the mechanism of ordinary post-antibiotic diarrhea, without any detectable intestinal mucosal lesion. It is certainly essential to obtain better understanding of why enterotoxinogenic *C. difficile* is not pathogenic in children under 2 years of age and why, on the contrary, it preferentially affects the elderly - role of propitious community environment, more frequent antibiotic treatment, debilitated underlying condition ?

As can be seen, many questions raised by digestive disorders due to *C. difficile* remain unanswered. The answer to many will be found in this work, thanks to the expertise of its contributors who have been good enough to share their knowledge with us. But, as always, answers will raise further questions, and what better could be hoped for to provide the answers than a third symposium in a few years time, thanks to the support of Biocodex Laboratories.

References

1. Bartlett JG (1988) Introduction. In: Rolfe RD, Finegold SM (eds) *Clostridium difficile*: its role in intestinal disease. Academic Press, San Diego, pp 1-13

2. Bartlett JG, Chang TW, Gurwith M, Gorbach SL, Onderdonk AB (1978) Antibiotic-associated pseudomembranous colitis due to toxin-producing Clostridia. N Engl J Med 298: 531-534

3. Bartlett JG, Onderdonk AB, Cisneros RL, Kasper DL (1977) Clindamycin-associated colitis due to a toxin-producing species of *Clostridium* in hamsters. J Infect Dis 136: 701-705

4. Finney JMT (1893) Gastroenterostomy for cicatrizing ulcer of the pylorus. Bull Johns Hopkins Hosp 4: 53-55

5. Green RH (1974) The association of viral activation with penicillin toxicity in guinea pigs and hamsters. Yale J Biol Med 3: 166-181

6. Hafiz S, McEntegart MG, Morton RS, Waitkins SA (1975) *Clostridium difficile* in the urogenital tract of males and females. Lancet i: 420-421

7. Hall IC, O'Toole E (1935) Intestinal flora in new-born infants. Am J Dis Child 49: 390-402

8. Hambre DM, Raki G, McKnee CM, Mac Phillanny HB (1943) The toxicity of penicillin as prepared for clinical use. Am J Med Sci 206: 642-653

9. Hummel RP, Altemeier WA, Hill EO (1984) Iatrogenic staphylococcal entero-colitis. Ann Surg 160: 551

10. Larson HE, Parry JV, Price AB, Davis DK, Bolly J, Tyrrel DA (1977) Undescribed toxin in pseudomembranous colitis. Br Med J 1: 1246-1248

11. Pettet JD, Baggenstoss AH, Dearing WH, Judd ES (1954) Postoperative pseudo-membranous colitis. Surg Gynecol Obstet 8: 547

12. Small JD (1968) Fatal enterocolitis in hamsters given lincomycin hydrochloride. Lab Anim Care 18: 411-420

13. Snyder ML (1940) The normal fecal flora of infants between two weeks and one year of age. J Infect Dis 66: 1-16

14. Tedesco FJ, Barton RW, Alpers DH (1974) Clindamycin-associated colitis. Ann Intern Med 81: 429-433

15. Wrod JS, Bennet IL, Yardley JH (1956) Staphylococcal enterocolitis in chinchillas. Bull Johns Hopkins Hosp 98: 454-463

Clostridium difficile: clinical spectrum with emphasis on atypical clinical presentations

Philippe Marteau

SUMMARY

The diagnosis of pseudomembranous colitis (PMC) is often easy. In fact, the presence of symptoms of frank colitis associated with systemic problems, with a history of recent antibiotic treatment, should prompt the need for an immediate colonoscopy, which will reveal the typical «false membranes». *C. difficile* and its toxins will be recovered from the stools in over 90% of cases. Nonetheless, the diagnosis of *C. difficile* colitis is at times difficult, due to atypical history, symptoms, lesions or microbiological results. Also there may be no recent antibiotic treatment. An associated disease such as inflammatory bowel disease or AIDS may be present. Rectal involvement is by no means invariable. «Pseudo-surgical» forms exist, such as fulminating colitis or toxic dilatation. The history and ultrasound results may be helpful, but the key investigation is endoscopy. Rare forms such as small bowel involvement, intra-abdominal abscesses or arthritis should be borne in mind. Stage III lesions with large confluent lesions, or at the other end of the scale early false membranes, or non pseudomembranous lesions, may give rise to diagnostic difficulty, in which case the histology and microbiology are of great value, though the latter may be difficult if toxin production is absent. Recurrences occur in some 10% to 20% of cases, and these tend to be repeated. Finally, the causal role of *C. difficile* in simple antibiotic related diarrhea is hard to establish, especially bearing in mind the existence of healthy carriers.

Clostridium difficile : diversité des aspects cliniques et notamment des présentations atypiques

RÉSUMÉ

Le diagnostic de colite pseudomembraneuse ne présente le plus souvent aucune difficulté. La présence de symptômes de colite franche, associée à des signes généraux et à une notion d'antibiothérapie récente doit rapidement conduire à la colonoscopie, qui révélera l'aspect typique de « fausses membranes ». *C. difficile* et ses toxines sont isolés dans les selles dans plus de 90 % des cas. Néanmoins, le diagnostic de colite à *C. difficile* est parfois plus complexe, en raison d'antécédents, de symptômes, de lésions ou de résultats microbiologiques atypiques. Ainsi, il peut ne pas y avoir de traitement antibiotique récent. Des pathologies telles qu'une maladie inflammatoire de l'intestin ou un SIDA peuvent être associées. L'atteinte du rectum, si elle est classique, est loin d'être systématique. On peut observer des formes chirurgicales, telles que la colite fulminante ou la dilatation toxique. L'histoire de la maladie et les résultats échographiques peuvent apporter une aide, mais l'investigation-clef est l'endoscopie. La possibilité de formes rares telles qu'une atteinte de l'intestin grêle, des abcès intra-abdominaux ou des arthrites doit être gardée en mémoire. Les formes de stade III avec de grandes lésions confluantes ou, à l'autre extrémité de l'échelle, les fausses membranes précoces et les lésions sans pseudomembrane, sont elles aussi à l'origine de difficultés diagnostiques. Dans ces cas, l'histologie et la microbiologie sont de grande valeur, bien que la dernière soit plus aléatoire en l'absence de production de toxines. Les rechutes surviennent dans 10 à 20 % des cas, et ont tendance à se répéter. Au total, le rôle étiologique de *C. difficile* dans la diarrhée simple aux antibiotiques est difficile à établir, surtout du fait de l'existence des porteurs sains.

Clostridium difficile: Klinisches Spektrum unter besonderer Berücksichtigung atypischer klinischer Erscheinungsformen

ZUSAMMENFASSUNG

In vielen Fällen ist die Diagnose einer pseudomembranösen Kolitis (PMK) einfach. Bei Vorliegen von Symptomen einer manifesten Kolitis mit systemischen Erscheinungen sollte bei Patienten, die kurz zuvor Antibiotika eingenommen haben, möglichst umgehend eine Koloskopie erfolgen, bei der die typischen "Pseudomembranen" sichtbar werden. In über 90 % der Fälle werden *C. difficile* und die von ihm produzierten Toxine im Stuhl nachgewiesen. Dennoch ist die Diagnose einer *C. difficile*-Kolitis bei atypischer Vorgeschichte, Symptomatik, Schädigung oder ungewöhnlichem mikrobiologischem Bild häufig problematisch: Möglicherweise ist zuvor gar keine Antibiotikabehandlung erfolgt, es können Begleiterkrankungen wie entzündliche Darmkrankheiten oder AIDS vorliegen. Auch ist nicht zwingend das Rektum betroffen. Es gibt "pseudochirurgische" Formen wie die fulminante Kolitis oder das toxische Megakolon. Anamnese und Ultraschalluntersuchungen können zwar hilfreich sein, ausschlaggebend ist jedoch die Endoskopie. Seltene Formen mit Dünndarmbeteiligung, intraabdominellen Abszessen oder Arthritis sollten berücksichtigt werden. Läsionen 3. Grades mit großen, konfluierenden Läsionsflächen oder andererseits ganz frühe Pseudomembranen oder das völlige Fehlen von pseudomembranösen Läsionen können die Diagnose ebenfalls erschweren. In solchen Fällen sind histologische und mikrobiologische Untersuchungen sinnvoll, wenn auch letztere bei fehlender Toxinproduktion unter Umständen schwierig sind. Zu Rezidiven kommt es in rund 10-20 % der Fälle, und diese neigen zu weiteren Rezidiven. Darüber hinaus ist die kausale Beteiligung von *C. difficile* an einfachen Antibiotika-assoziierten Diarrhöen schwer nachweisbar, vor allem, wenn man berücksichtigt, daß der Keim auch bei gesunden Trägern nachweisbar ist.

Introduction

The classical clinical picture and the one which most easily incriminates *C. difficile* is that of pseudomembranous colitis (PMC), promoted by antibiotics. Indeed: (1) almost all cases of PMC are associated with *C. difficile*; (2) almost all cases of PMC follow recent antibiotic treatment; (3) eradication of *C. difficile* cures PMC and (4) *C. difficile* can induce very similar illnesses in animal models [14]. To impute other clinical conditions to *C. difficile* is a much more difficult task. The clinician should be on his guard to remember the role of *C. difficile* when faced with unusual signs or lesions, but must also bear in mind the existence of healthy carriers and the futility and potential dangers of attempting to treat them [10].

Typical stage II pseudomembranous colitis

This should not raise diagnostic difficulties when it occurs during or after treatment with antibiotics, or in the presence of other risk factors such as chemotherapy. Here, there are frank signs of colitis, often combined with serious systemic effects [15, 26]. They should prompt an immediate endoscopy, which will reveal the false membranes. Additionally, *C. difficile* and its toxins will be recovered from the stools in over 90% of cases.

Symptoms

PMC presents in 95% of cases with diarrhea. Typically, this is profuse and liquid, and results in the passage of 4 to 10 creamy or greenish stools per day [15, 26]. Colic occurs in 70% of cases and fever is often present (above 37.5°C in 66% of cases, and over 38.5°C in 26%). Signs of dehydration are seen in 30%, and weight loss of over 10% body weight in 23% [1]. Bleeding occurs in less than 5%. There is often a leucocytosis (above 15,000/ml in 40% of cases and over 20,000 in 20%). Hypoalbuminaemia is common, below 25g/l in 24% of cases [15, 16]. Hypo-cholesterolaemia occurs in half the cases, and a protein losing enteropathy is invariable [19]. Blood cultures are usually negative.

Diagnosis

A history of recent treatment with antibiotics is present in 95% of cases. The average time from the beginning of treatment to the appearance of symptoms is 7 days, although periods of as long as 10 weeks have been recorded, and quite often the symptoms begin only several days after treatment has been discontinued. While suspected on the clinical findings and history of antibiotic treatment, the diagnosis must be confirmed by investigations. The rectum is involved in over 70% of cases [28], and the lesions may be felt on digital examination or seen on sigmoidoscopy. Endoscopy reveals slightly elevated yellowish false membranes, of from 2 to 20 mm across, which may be diffuse or localised. Histologically, they are composed of the necrotic superficial half of the colonic mucosa with an exudate of fibrin and leucocytes, tissue debris and mucus [13]. The presence of mucus, which stains well with Alcian blue, is an important element in the differential diagnosis from other forms of colitis.

The underlying mucosa contains crypts distended with abundant mucus and an inflammatory infiltrate of the lamina propria containing many polymorphs, and the submucosa is often edematous. The stools contain *C. difficile*, and tests for the toxins are positive.

Other investigations are of less value. The barium enema demonstrates non specific thumbprints, and computed tomography (CT) shows thickening of the bowel wall (usually localised to the rectum) in 61% of cases. Exaggerated protrusions in an «accordion» pattern are seen in 5 to 17% and clinically undetectable ascites in 15%. The sensitivity of CT lies at around 85% and the specificity at 48% [4].

With appropriate treatment, the clinical course shows resolution of the general signs within 1 to 3 days, with disappearance of diarrhea in 5 days. Recurrence is frequent (see below). Without treatment, the course worsens in those cases where the antibiotic treatment is continued, but even if antibiotics are withdrawn, spontaneous improvement is seen in less than 25% of cases.

Stage I and III pseudomembranous colitis are more difficult to diagnose. In patients with a severe illness resembling pseudomembranous colitis, endoscopic lesions may be minimal. Thus, early false membranes of 1 to 2 mm across (the so-called stage I PMC) may be difficult to see, the biopsies then assume particular importance and may show a focal necrosis of the superficial epithelium combined with a discrete fibrino-leucocytic exudate [13]. The discovery of focal hyperplasia of the superficial epithelium combined with polymorph exocytosis (described in experimental pathology) may also help in the diagnosis.

At the opposite, forms of the disease with major colonic ulcerative lesions, the so-called stage III of the disease may be impossible to distinguish either endo-scopically or on histology from ischemic, infective or inflammatory colitis. The finding of toxinogenic strains of *C. difficile* is helpful, but not diagnostic, because it can on rare occasions be present in other forms of inflammatory colitis, infections or parasites. These difficult situations need careful testing of the stools for other bacteria and parasites in order to exclude these other forms and very careful histological study of colonic and rectal biopsies may also be required. The situation is often not completely clarified until after several days of treatment.

PMC without diarrhea, or with ileus or toxic megacolon, and fulminating types of disease

True PMC may occur with no diarrhea [5, 6, 8, 11, 15, 30] especially after surgery. The clinical picture of the «pseudo-surgical» acute abdomen should be remembered. Systemic symptoms are usually severe, with high fever, tachycardia, and sometimes shock, intense abdominal pain is common. Toxic megacolon and paralytic ileus may occur. Abdominal examination shows meteorism, and tenderness and guarding are seen as in appendicitis or peritonitis. Minor degrees of ascites or pleural effusion may be present. The diagnosis should be considered

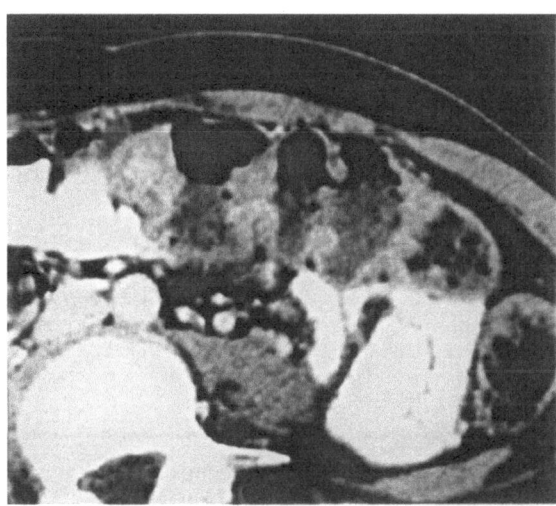

Fig. 1.
CT appearances in a patient with pseudomembranous colitis: the « accordion sign »

when there is a recent history of treatment with antibiotics or where there is a suggestive scan (Fig. 1). The diagnosis is often made at laparotomy, which reveals severe edema of the colonic wall [21]. The best method is certainly an urgent colonoscopy, which should always be done if there is no sign of perforation [30]. Initial treatment consists of decompression via the colonoscope, restoration of fluid and electrolyte balance and metronidazole, given either orally or through a naso-gastric tube or through a decompression tube inserted in the colon or intravenously (250 mg every 6 hours), or vancomycin. If there is no response within 48 to 72 hours, most authors would recommend subtotal colectomy and temporary ileostomy [5, 11].

Small bowel involvement

Several authors have reported the presence of *C. difficile* in the small intestine, in the absence of lesions [29]. The role of the toxins of *C. difficile* was discussed in one case of severe diarrhea through an ileostomy [23]. A few rare cases of small bowel involvement with false membrane formation both in the colon and small bowel have been recorded [13], as well as a case of ileal ulceration with gas in the gut wall [12]. There has even been a report of an ileal urinary conduit having been affected [24].

C. difficile colitis without PMC
and *C. difficile* diarrhea without colitis

Pseudomembranous colitis represents only one, albeit the most serious, of the effects which *Clostridium difficile* produces in the gut [11, 15]. The other

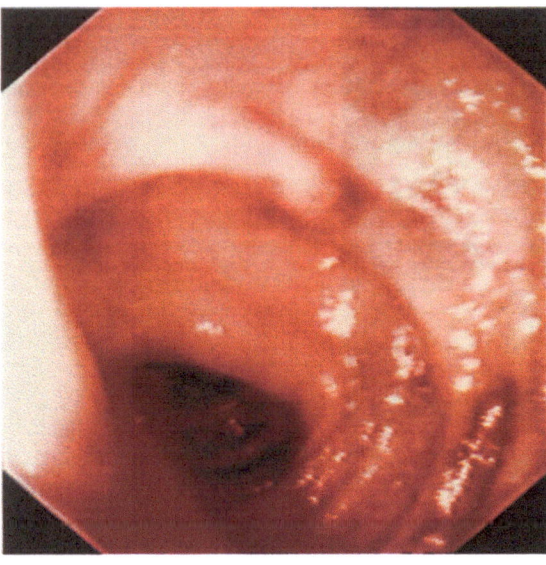

Fig. 2.
Erosive colitis due
to *Clostridium difficile*
(by courtesy of Dr Jean-Luc
Larpent, Clermont-Ferrand)

syndromes are less well known, for three reasons: (1) they are very variable and clinicians do not always have the facilities to identify *Clostridium difficile* and its toxins in the stools; (2) the differential diagnosis lies between infectious and ischemic colitis; (3) given the existence of healthy carriers, the clinician must always exclude the possibility of another form of colitis when toxinogenic *C. difficile* is found in the stools [31].

C. difficile seems to be responsible for some 20% to 30% of all diarrhea associated with antibiotics, in the absence of PMC. The organism can be recovered from the stools in 50% to 70% of cases of non-PMC colitis (Fig. 2) following antibiotics, and in 15% to 25% of cases of simple diarrhea with no detectable colonic lesion. Colitic changes are more frequently found in patients who present with alimentary or systemic symptoms [26].

In one series of 98 patients carrying *C. difficile*, Talbot et al found that where the symptoms were severe there was a 71% incidence of PMC and 21% of other types of colitis, with moderate symptoms the figures were respectively 35% and 19%, and with mild symptoms 23% and 23% [26]. Non specific lesions can be segmental or diffuse and they may include erythema, edema, ulceration, or more rarely hemorrhage [11, 15, 26]. Biopsies show a polymorph infiltrate between the crypts [13].

It is very difficult to incriminate *C. difficile* in simple antibiotic related diarrhea or in acute diarrhea with no obvious etiology. In effect, in patients receiving antibiotics, one recovers toxin secreting *C. difficile* almost as often from the stools of those without diarrhea (15-30% of cases) as from those with «simple» diarrhea. This type of condition does not as a rule respond to antibiotic treatment designed to eradicate *C. difficile*.

Healthy carriers of *C. difficile*

From one to two percent of western adults treated with antibiotics and nearly 70% of infants are healthy carriers of *C. difficile* [3, 14]. Treatment with vancomycin or metronidazole seems not to be effective in that if clearance occurs it is usually transitory and carries a greater risk of reinfection than does a placebo [10].

C. difficile and other alimentary disorders

It is no longer thought that *Clostridium difficile* has a causative role in inflammatory bowel diseases of the intestine, or promotes the attacks [3, 11, 15]. The clinician must however bear in mind that pseudomembranous colitis can supervene in a patient with inflammatory colitis, especially (but not necessarily) following treatment with antibiotics. The hypothesis that assigns to *C. difficile* a causative role to neonatal necrotising enterocolitis or the irritable bowel syndrome has been abandoned [15].

Recurrent forms

Whereas antibiotic treatment is often effective in the short term in the treatment of *C. difficile* infections, recurrence nonetheless occurs in 10% to 20% of cases and in these patients a further recurrence is seen in 40% of cases [3, 11, 14]. These recurrent cases represent an important subgroup for therapeutic studies. Recurrence is due as much to the re-emergence of the same strain of *C. difficile* (particularly since they are spore forming) as to reinfection by a different strain and is probably due to the persistence of depressed flora antagonising *C. difficile* and/or to specific immune deficiency.

Systemic manifestations

Reactive arthritis

Eighteen cases of arthritis associated with *C. difficile* have been reported [17, 22]. These usually involve a limited arthritis coming on 18 days after the beginning of the diarrhea (but sometimes in the absence of any diarrhea). A fever of 39°C is usual, the leucocyte count averages of 12,000/ml, the erythrocyte sedimentation rate is raised. Rheumatoid factors are absent, antinuclear antibodies are present in 60% of cases in weak concentration and 80% of males and 44% of females carry HLA-B27 antigens [17]. Three of these patients presented with Reiter's syndrome. Cure was obtained at an average period of six months and the long term prognosis appears excellent.

Systemic infections and abscesses

C. difficile is a non invasive organism. However some rare examples of pancreatic, splenic or intra-abdominal abscesses have been reported, as well as one case of *C. difficile* osteomyelitis [18, 20, 25].

C. difficile diarrhea and colitis complicating AIDS

C. difficile may be responsible for 4% to 12% of cases of diarrhea occurring in the course of AIDS infection [9]. The risk is increased by antibiotics to which these patients are frequently exposed, by the length of hospital stay (particularly in units colonised by *C. difficile*), but apparently not by the extent of depletion of CD4 lymphocytes [2, 9]. Treatment with rifabutine or antiviral agents have also been identified as possible risk factors.

The symptoms and signs are similar to those of usual PMC occurring in an immunocompetent subject. In one series of 17 AIDS patients who were infected by *C. difficile* the diarrhea resolved in 80% with treatment [7]. This suggests that *C. difficile* diarrhea is as responsive to appropriate treatment as in the patients who are not infected with the virus, and that *C. difficile* is a treatable cause of diarrhea in AIDS patients, and should therefore be carefully looked for.

Conclusions

The clinician should bear in mind the possible role of *C. difficile* when faced with bowel symptoms with or without systemic involvement, in a situation which favors its development, such as recent treatment with antibiotics. A search should be made for the organism and its toxins in the stools, and the typical colonic lesions sought. This allows a confident and accurate diagnosis of PMC, whether or not the presentation is typical, or another condition is present. Rarely, the causative role of *C. difficile* may be in doubt, if the symptoms are mild and the colonic lesions atypical or absent. The clinician must then decide on whether the presence of *C. difficile* is incidental or pathogenic, and take the appropriate therapeutic measures.

References

1. Anand A, Bashey B, Tanveer M, Glatt AE (1994) Epidemiology, clinical manifestations, and outcome of *C. difficile*-associated diarrhea. Am J Gastroenterol 89: 519-523
2. Barbut F, Mario N, Meyohas MC, Binet D, Frottier J, Petit JC (1994) Investigation of a nosocomial outbreak of *C. difficile*-associated diarrhea among AIDS patients by random amplified polymorphic DNA (RAPD) assay. J Hosp Infect 26: 181-189

3. Bartlett JG (1994) *C. difficile*: history of its role as an enteric pathogen and the current state of knowledge about this organism. Clin Infect Dis 18 (suppl 4): S265-272

4. Boland GW, Lee MJ, Cats AM, Gaa JA, Saini S, Mueller PR (1994) Antibiotic-induced diarrhea: specificity of abdominal CT for the diagnosis of *C. difficile* disease. Radiology 191: 103-106

5. Bradley SJ, Weaver DW, Maxwell NPT, Bouwman DL (1988) Surgical management of pseudomembranous colitis. Am Surg 54: 329- 332

6. Cone JB, Wetzel W (1982) Toxic megacolon secondary to pseudomembranous colitis. Dis Colon Rectum 25: 478-482

7. Cozart JC, Kalangi SS, Clench MH, Taylor DR, Borucki MJ, Pollard RB, Soloway RD (1993) *C. difficile* diarrhea in patients with AIDS versus non-AIDS controls. J Clin Gastroenterol 16: 192-194

8. Drapkin MS, Worthington MG, Chang TW, Razvi SA (1985) *C. difficile* colitis mimicking acute peritonitis. Arch Surg 120: 1321-1322

9. Hutin Y, Molina JM, Casin I, Daix V, Sednaoui P, Welker Y, Lagrange P, Decazes JM, Modaï J (1993) Risk factors for *C. difficile*-associated diarrhoea in HIV-infected patients. AIDS 7: 1441-1447

10. Johnson S, Homann SR, Bettin KM, Quick JN, Clabots CR, Peterson LR, Gerding DN (1992) Treatment of asymptomatic *C. difficile* carriers (fecal excretors) with vancomycin or metronidazole. Ann Intern Med 117: 297-302

11. Kelly CP, Pothoulakis C, LaMont JT (1994) *C. difficile* colitis. N Engl J Med 330: 257-262

12. Kuntz DP, Shortsleeve MJ, Kantowitz PA, Gauvin GP (1993) *C. difficile* enteritis: a cause of intramural gas. Dig Dis Sci 10: 1942-1944

13. Lavergne A, Galian A (1990) *Clostridium difficile* et anatomie pathologique. In: Rambaud JC, Ducluzeau R (eds) *Clostridium difficile* et pathologie intestinale. Springer Verlag, Paris, pp 1-16

14. Marteau P, Sobhani I, Berretta O, Rambaud JC (1991) Physiopathologie des infections intestinales dues à *Clostridium difficile*. Rôle de l'écosystème colique. Gastroenterol Clin Biol 15: 322-329

15. Matuchansky C (1990) Spectre clinique des infections intestinales à *Clostridium difficile*. In: Rambaud JC, Ducluzeau R (eds) *Clostridium difficile* et pathologie intestinale. Springer Verlag, Paris, pp 17-23

16. Mogg GAG, Keighley MRB, Burdon DW, et al (1979) Antibiotic-associated colitis. A review of 66 cases. Br J Surg 66: 738-742

17. Putterman C, Rubinow A (1993) Reactive arthritis associated with *C. difficile* pseudomembranous colitis. Semin Arthritis Rheum 22: 420-426

18. Riley TV, Karthigasu KT (1982) Chronic osteomyelitis due to *C. difficile*. Br Med J 284: 1217-1218

19. Rybolt AH, Laughon BE, Greenough III WB, Bennett RG, Thomas DR, Bartlett JG (1989) Protein-losing enteropathy associated with *Clostridium difficile* infection. Lancet i: 1353-1355

20. Sagimur R, Fogel R, Begin L, Cohen B, Mendelson J (1983) Splenic abcess due to *C. difficile*. J Infect Dis 147: 1105

21. Schnitt SJ, Antonioli DA, Goldman H (1983) Massive intramural edema in severe pseudomembranous colitis. Arch Pathol Lab Med 107: 211-213

22. Sensini A, Marroni M, Bassotti G, Farinelli S, D'Alo F, Gentili AM, Sbaraglia G, Baldelli F (1993) *C. difficile*-associated reactive arthritis in an HLA-B27 negative male. J Clin Gastroenterol 16: 354-358

23. Sharpstone D, Hudson MJ, Colin-Jones DG (1993) *C. difficile*-induced small bowel diarrhea. Eur J Gastroenterol Hepatol 5: 759-760

24. Shortland JR, Spencer RC, Williams JL (1983) Pseudomembranous colitis associated with changes in an ileal conduit. J Clin Pathol 36: 1184-1187

25. Sofianou DC (1988) Pancreatic abscess caused by *Clostridium difficile*. Eur J Clin Microbiol Infect Dis 7: 528-529

26. Talbot RW, Walker RC, Beart RW (1986) Changing epidemiology, diagnosis and treatment of *C. difficile* toxin-associated colitis. Br J Surg 73: 457-460

27. Taylor R, Borriello SP, Taylor AJ (1981) Isolation of *C. difficile* from the small bowel. Br Med J 283: 412

28. Tedesco FJ (1979) Antibiotic associated pseudomembranous colitis with negative proctosigmoidoscopic examination. Gastroenterology 77: 225-227

29. Testore GP, Nardi F, Babudieri S, Giuliano M, DiRosa R, Panichi G (1986) Isolation of *C. difficile* from human jejunum: identification of a reservoir for disease? J Clin Pathol 39: 861-862

30. Triadafilopoulos G, Hallstone AE (1991) Acute abdomen as the first presentation of pseudomembranous colitis. Gastroenterology 101: 685-691

31. Viscidi R, Willey S, Bartlett JG (1981) Isolation rates and toxigenic potential of *C. difficile* isolates from various patient populations. Gastroenterology 81: 5-9

Clostridium difficile infection in the elderly

Richard G. Bennett

SUMMARY

Clostridium difficile infection is likely to become an increasing problem in clinical practice as the number of older patients in the population grows. Although the prevalence of asymptomatic carriage does not appear to increase across the lifespan, the incidence of symptomatic disease increases dramatically from the third to the eighth decade of life, and age is regularly identified as a risk factor for infection. As with younger adults, most older patients develop *C. difficile* diarrhea during or immediately following treatment with systemic antibiotics. However, unusual manifestations of illness which may be more common among older as compared to younger patients include high fever (even 39.5-40.5°C in the elderly), toxic megacolon (colitis without diarrhea), leukemoid reactions (white blood cell counts of 25,000-50,000/mm^3), and multiple recurrent episodes of diarrheal illness. Severe and recurrent disease can be particularly problematic in the elderly who are at high risk for malnourishment generally, because infection results in protein-losing enteropathy which can worsen the nutritional state.

Outbreaks of *C. difficile* have been reported in a number of long-term care facilities and nursing homes. However, *C. difficile* is probably not endemic at high levels of colonization among residents of these types of facilities generally, and true outbreaks with rapid spread are probably rare. Instead, the more usual scenario is that low levels of colonization are maintained as patients with unrecognized carriage are discharged from hospitals to nursing homes regularly, and new cases arise given the frequent use of antibiotics in these facilities, and because crowding and fecal contamination facilitate spread of enteric infections. In addition, in many facilities with a perceived *C. difficile* problem, increased vigilance rather than spread of infection no doubt underlies, at least in part, the identification of high numbers of cases. In all long-term care settings, a rational approach is recommended with regards to an infection control program.

The most crucial components of any program are: (1) regular in-service training in hand-washing; (2) maintenance of a high index of suspicion for this infection; and (3) rational use of antibiotics.

Avoidance of the prescription of antibiotic treatment for elderly patients whenever possible is a goal to minimize the incidence of *C. difficile* disease. In elderly patients on antimicrobial therapy who do develop diarrhea, the antibiotic should be discontinued if possible. When diarrhea occurs, attention should be paid to maintaining hydration because underlying atherosclerosis in older patients can lead to infarctions of vital organs with volume depletion.

Oral rehydration therapy should be the mainstay of treatment in this regard. Because antiperistaltic drugs can precipitate the development of toxic megacolon in individuals with *C. difficile* infection, these drugs should be avoided absolutely in patients at risk for this infection. Bismuth subsalicylate is a beneficial adjunct treatment for patients with diarrhea generally, can be given safely to the elderly, and is recommended when *C. difficile* disease is suspected. (Bismuth subsalicylate has *in vitro* activity against *C. difficile*, and is protective in an animal model of this disease.) Metronidazole and vancomycin remain the mainstays of antimicrobial therapy for patients with moderate and severe disease, respectively, and weeks of treatment are sometimes required for patients with relapsing *C. difficile* infection. For the future, microbial replacement therapy holds promise, particularly for treating elderly patients with recurrent disease.

L'infection à *Clostridium difficile* chez le sujet âgé

RÉSUMÉ

Les infections à *C. difficile* deviennent un problème de plus en plus fréquent en pratique clinique, en raison du nombre de personnes âgées qui augmente dans la population. La prévalence du portage asymptomatique ne semble pas augmenter au cours de la vie, mais l'incidence des formes symptomatiques de la maladie s'élève considérablement de la 3ème à la 8ème décennie, et l'âge est régulièrement identifié comme un facteur de risque d'infection à *C. difficile*. Comme les adultes jeunes, la plupart des patients âgés développent une diarrhée à *C. difficile* pendant ou au décours d'un traitement antibiotique par voie générale. Cependant, les manifestations inhabituelles de la maladie, qui incluent une fièvre élevée (39,5°C à 40,5°C chez le sujet âgé), un mégacôlon toxique (colite sans diarrhée), des réactions de type leucémique (nombre de globules blancs compris entre 25 et 50.000/mm^3) et des épisodes de diarrhée multirécidivante, sont plus fréquentes parmi les sujets âgés comparativement aux patients plus jeunes. Une maladie sévère et récidivante peut être particulièrement problématique chez les sujets âgés qui présentent souvent un risque élevé de malnutrition. En effet, l'infection se traduit par une entéropathie avec perte protéique, qui peut aggraver l'état nutritionnel.

Des épidémies à *C. difficile* ont été plusieurs fois décrites dans des établissements de long séjour et des centres de soins. Cependant, la présence de *C. difficile* à des niveaux élevés de colonisation n'est probablement pas endémique parmi les résidents de ce type d'établissement, et de véritables épidémies avec une propagation rapide sont sans doute rares. En fait, le scénario le plus habituel est celui de bas niveaux de colonisation, maintenus par des porteurs asymptomatiques, et régulièrement transmis des hôpitaux vers les centres de soins. De plus, l'émergence de nouveaux cas est favorisée par l'utilisation fréquente des antibiotiques dans ces centres, mais aussi par la promiscuité et la contamination fécale qui facilitent la propagation des infections intestinales. Dans de nombreux établissements avec un problème reconnu d'infection à *C. difficile*, c'est une vigilance accrue plutôt que la dissémination du germe qui explique, au moins en partie, l'identification d'un nombre élevé de cas. Dans tous les centres de long séjour, une approche rationnelle est recommandée, comportant en particulier un programme de contrôle de l'infection.

Les principaux éléments de tout programme de ce type sont : (1) la pratique régulière du lavage des mains dans le service; (2) un degré élevé de suspicion pour cette infection; (3) l'usage rationnel des antibiotiques.

Éviter, dans la mesure du possible, la prescription d'un traitement antibiotique chez les sujets âgés constitue une mesure importante pour réduire l'incidence des infections à *C. difficile*.

Chez les patients âgés sous antibiotique et qui développent une diarrhée, le traitement en cause doit si possible être arrêté. En cas de diarrhée, l'attention doit être portée sur le maintien de l'hydratation, car l'athérosclérose, sous-jacente chez les patients âgés, peut conduire à un infarctus des organes vitaux en cas de déplétion volumique. Dans cette optique, la réhydratation orale doit être le point essentiel du traitement. Les antipéristaltiques peuvent précipiter le développement d'un mégacôlon toxique chez les individus porteurs d'une infection à *C. difficile*, et ces médicaments doivent absolument être évités chez les patients susceptibles de présenter cette pathologie. Le subsalicylate de bismuth constitue un traitement adjuvant possible pour les patients atteints d'une diarrhée (il peut être donné aux sujets âgés avec une bonne sécurité d'emploi et est recommandé lorsqu'une infection à *C. difficile* est suspectée, car il possède une activité *in vitro* contre *C. difficile* et un rôle protecteur dans les modèles animaux de cette pathologie). Le métronidazole et la vancomycine demeurent les éléments essentiels de la thérapeutique antibiotique chez les patients porteurs d'une infection à *C. difficile* modérée ou sévère, et plusieurs semaines de traitement sont parfois nécessaires pour les patients présentant des récidives. Les thérapies microbiennes de substitution sont des techniques d'avenir, qui d'ores et déjà tiennent leurs promesses, particulièrement pour le traitement de sujets âgés atteints de formes récidivantes.

Clostridium difficile-Infektionen bei älteren Patienten

ZUSAMMENFASSUNG

C. difficile-Infektionen dürften mit dem steigendem Anteil älterer Menschen an der Gesamtbevölkerung zunehmend zu einem Problem in der klinischen Praxis werden. Auch wenn die Prävalenz asymptomatischer Infektionen mit diesem Keim im Laufe des Lebens offenbar nicht zunimmt, steigt die Inzidenz einer symptomatischen Erkrankung zwischen dem dritten und dem achten Lebensjahrzehnt deutlich an, und ein höheres Alter wird in vielen Fällen als Risikofaktor für eine Infektion ermittelt. Ebenso wie bei jüngeren Erwachsenen treten *C. difficile*-bedingte Diarrhöen auch bei den meisten älteren Patienten während oder unmittelbar nach einer Therapie mit systemisch wirkenden Antibiotika auf. Allerdings kommt es bei älteren Patienten offenbar häufiger als bei jüngeren zu ungewöhnlichen Krankheitsbildern mit hohem Fieber (bis zu 39,5-40,5 °C bei Senioren), toxischem Megakolon (Kolitis ohne Diarrhö), leukämoiden Reaktionen (Leukozytenzahlen von 25.000-50.000/mm^3) sowie multipel rezidivierenden Diarrhöen. Schwere und rezidivierende Erkrankungen sind besonders bei älteren Patienten problematisch, bei denen ohnehin ein generell erhöhtes Risiko einer Mangelernährung besteht, da die Infektion eine Enteropathie mit Eiweißverlusten bedingt und so den Ernährungszustand weiter verschlechtern kann.

Aus zahlreichen Langzeit-Pflegeeinrichtungen und -heimen sind Häufungen von *C. difficile*-Infektionen bekannt. Dennoch scheint *C. difficile* bei Bewohnern solcher Heime im allgemeinen nicht in Form einer hochgradigen Besiedelung endemisch zu sein. Zu echten, sich rasch unter den Heimbewohnern ausbreitenden Epidemien kommt es offenbar nur selten. Statt dessen besteht üblicherweise ständig eine geringgradige Durchseuchung, zum einen, da regelmäßig Patienten mit unerkannter Infektion aus Krankenhäusern in Pflegeheime überwiesen werden, bedingt durch die verbreitete Verwendung von Antibiotika in solchen Einrichtungen Neuerkrankungen hinzukommen und die Ansteckung mit Darminfektionen durch die Überfüllung der Heime und fäkale Kontamination begünstigt wird. Darüber hinaus ist bei vielen Einrichtungen, die Probleme mit *C. difficile* festgestellt haben, die angegebene hohe Fallzahl zumindest teilweise eher durch eine bessere Überwachung als durch eine stärkere Ausbreitung der Infektion bedingt. Langzeit-Pflegeeinrichtungen ist generell ein vernünftiges Vorgehen im Sinne eines Infektionsverhütungsprogramms zu empfehlen.

Die wichtigsten Elemente eines solchen Programms sind stets: (1) regelmäßige Schulung des Personals im Händewaschen, (2) Führen einer detaillierten Liste der Verdachtsmomente für diese Infektion und (3) vernünftiger Umgang mit Antibiotika.

Eine möglichst restriktive Antibiotikaverordnung an ältere Patienten ist ein geeignetes Mittel, um die Inzidenz von *C. difficile*-Erkrankungen deutlich zu senken. Kommt es bei älteren Patienten, die antimikrobiell behandelt werden, zu einer Diarrhö, sollte das Antibiotikum möglichst sofort abgesetzt werden. Bei Auftreten von Diarrhöen sollte auf eine Erhaltung des Flüssigkeitshaushaltes geachtet werden, da bei älteren Patienten eine Atherosklerose als Grunderkrankung Infarkte lebenswichtiger Organe mit Volumenmangel zur Folge haben kann.

Introduction

Clostridium difficile infection is diagnosed frequently in elderly patients treated in hospitals, and regularly encountered by clinicians caring for older individuals at home or in a nursing home. Since identified in the 1970's as the cause of almost all cases of pseudomembranous colitis, as well as the agent responsible for many cases of antibiotic-associated diarrhea [2, 12], this enteropathogen has become increasingly recognized as a significant problem in older adults. Most clinical laboratories rely upon tissue culture assays, enzyme immunoassays, and latex agglutination tests to identify *C. difficile* toxin(s) and diagnose infection. Although the tissue culture assay remains the « gold standard », a minimum of 48 hours is required before a result is reported as positive. Because of this temporal limitation, the immunologically-based tests are now ascendant.

In virtually every patient with pseudomembranous colitis confirmed endoscopically, toxin(s) can be identified in a stool sample. For patients with diarrhea during or following a course of antibiotics, only 15 to 25 percent will typically have a positive assay or test. Although few clinical laboratories rely upon stool culture for diagnosing *C. difficile* infection, whole stool or rectal swab cultures are regularly employed in research-based epidemiologic studies. It is critical to remember that culturing for *C. difficile* can routinely identify carriers of the organism among hospital and nursing home patients who will have negative assays or tests for toxin(s), and that some patients with diarrhea will have a negative assay or test for toxin(s), but a positive culture. Similarly, if in the future the identification of carriage relies upon amplification of toxin gene sequences [15], the epidemiology of this organism will need to be re-evaluated. These observations underlie the development of rational approaches to infection control within long-term care facilities which will be considered below.

Spectrum of *C. difficile* disease among the elderly

In the largest epidemiologic study of *C. difficile* published, investigators in Sweden determined that the carriage rate of *C. difficile* was 3 percent among 600 healthy volunteers who had not been treated with antibiotics within the previous 6 weeks [1]. In our own investigations of the prevalence of *C. difficile* across the lifespan, more than 700 specimens from participants in the Baltimore Longitudinal Study on Aging were analyzed. There was no age-related increased prevalence identified, as only four subjects had a positive stool culture and none had a positive toxin assay [R Bennett, unpublished data]. Thus, there is no current evidence that age alone is a risk factor for being asymptomatically colonized with this organism. Since *C. difficile* cannot be recovered from the vast majority of adults using current culture methodologies, a positive stool culture or toxin assay for *C. difficile* should always be regarded as abnormal regardless of the patient's age.

Although the risks of carriage have not been established, individuals who harbor this organism should intuitively be at higher risk for becoming symptomatic following exposure to antimicrobials, and should also be more likely to contribute to environmental contamination. More interesting, is the association which has been seen between carriage and subsequent mortality in the elderly. In two separate studies, an association between carriage and death was identified [3, 21]. However, since few of the subsequent deaths were obviously related to infection with *C. difficile*, carriage is most probably simply a marker for subsequent increased mortality, i.e., only a very rare patient who is known to be colonized with *C. difficile* appears to die from the infection.

In both younger and older adults, *C. difficile* diarrhea typically occurs only in individuals who have been treated with antibiotics, but the incidence increases with age. An analysis of almost 5,000 specimens from individuals with diarrhea in Sweden showed that the incidence of *C. difficile* diarrhea increased dramatically from less than 30 cases per million in the third decade of life to 250 cases per million in the eighth [1]. In another retrospective investigation of a hospital outbreak of *C. difficile* [9], and in a prospective study of *C. difficile* infection among hospitalized patients [18], age was also identified as a risk factor for developing infection. Since age does not appear to be a risk factor for carriage, other co-morbid conditions which predispose the elderly to infections generally, and especially the frequency with which antibiotics are prescribed to older patients, probably explain these observations.

Pseudomembranous colitis is a life-threatening infection caused by *C. difficile* which is classically characterized by diarrhea, high fever (even 39.5-40°C in elderly patients), and abdominal distension and pain. A high index of suspicion should be maintained for this diagnosis whenever an older patient has a fever without a probable source, or has a distended abdomen. In such patients, it is imperative to remember that pseudomembranous colitis can present not with diarrhea, but with marked abdominal distension and radiographic evidence of toxic megacolon (Fig. 1) resulting from dysmotility and/or obstruction from mucosal inflammation (Fig. 2). In addition, leukemoid reactions with white blood cell counts of 25,000 to 50,000/mm^3 can be seen in such cases. Considering the diagnosis of pseudomembranous colitis in these situations and beginning treatment promptly can prevent exploratory laparotomy or even death in an affected patient.

In older patients who are diagnosed with *C. difficile* diarrhea or pseudo-membranous colitis, the risk of a recurrent bout of diarrhea may be higher than in younger patients. Although only 5 to 10 percent of adults who develop symptomatic *C. difficile* infection will have a relapse, relapses among elderly nursing home patients are not unusual. Studies using typing methods have shown

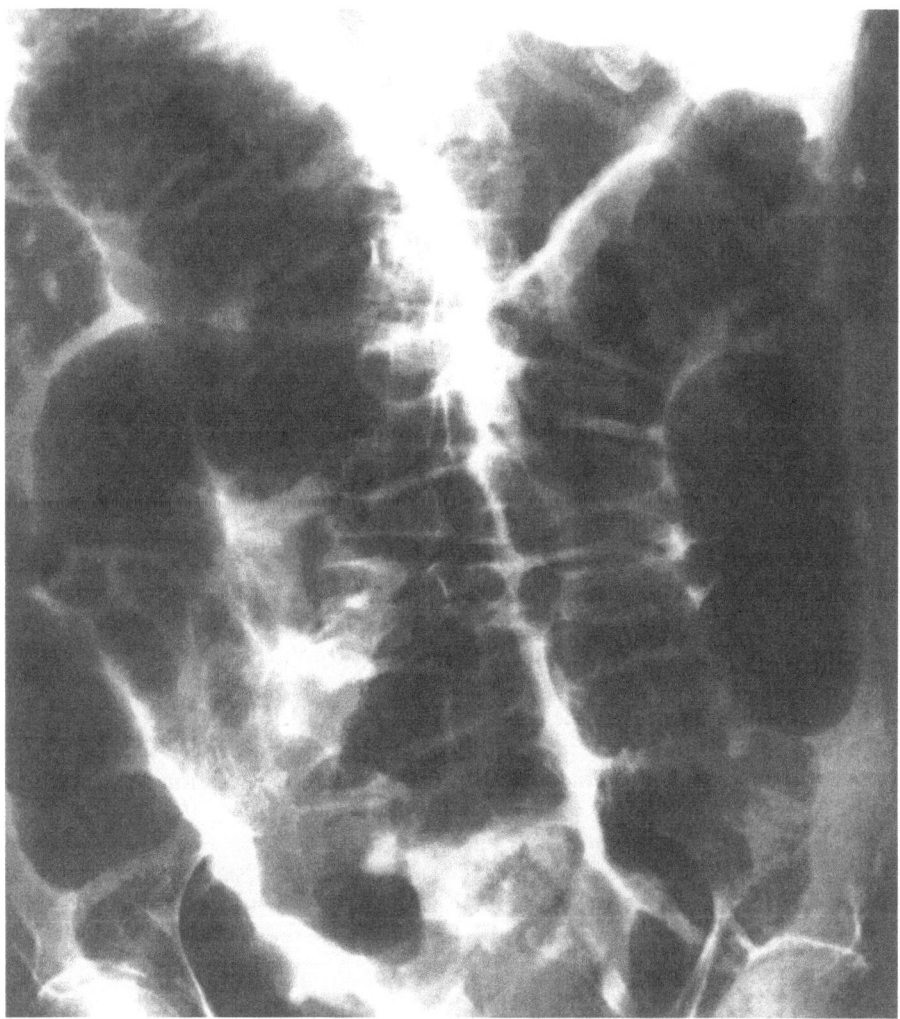

Fig. 1. An abdominal radiograph of a nursing home patient with a history of recurrent *Clostridium difficile* diarrhea who developed toxic megacolon after being treated with diphenoxylate hydrochloride with atropine sulfate. (Originally published by the author [4] and used with the permission of Geriatrics)

that recurrent diarrhea can result from the same organism, i.e., a true relapse, or from a different strain, i.e., re-infection leading to new symptoms [7, 17]. Unusual older patients may have many bouts of recurrent diarrhea over many months, and it is this group of patients who remain the most difficult to manage.

Finally, the elderly also may be at increased risk for the nutritional consequences of *C. difficile* infection. If serial bouts of diarrhea and fever occur, these recurrent illnesses can clearly worsen the nutritional state. In addition, *C. difficile*

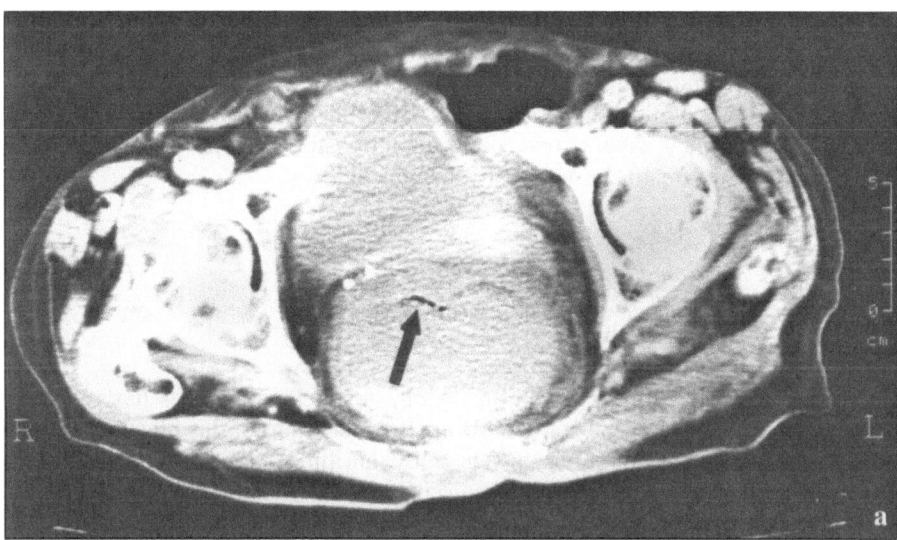

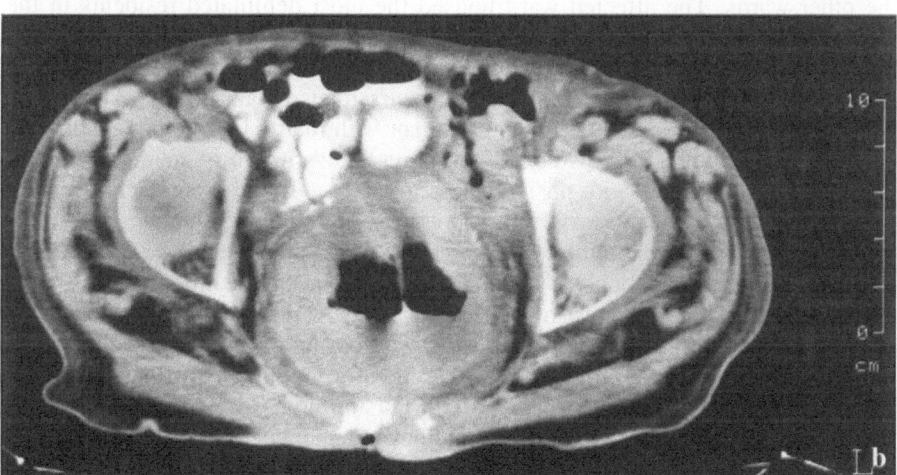

Fig. 2. Computed tomographs of the abdomen from the same patient as in Fig. 1 showing (**a**) an obstructive rectal mass with only minimal luminal gas (arrow) and (**b**) subsequent resolution of the obstruction five days later following treatment for pseudomembranous colitis with vancomycin. (Originally published by the author [4] and used with the permission of Geriatrics)

infection results in protein-losing enteropathy. Significant loss of serum proteins into the gut with pseudomembranous colitis can result in anasarca, and based upon the identification of alpha 1-antitrypsin* in stool specimens, protein-losing entero-

---

*Alpha 1-antitrypsin is a serum protein not usually found in the stool which is resistant to enzymatic and bacterial degradation. Presence of excess alpha 1-antitrypsin in stool is a marker for protein-losing enteropathy.

pathy also occurs in patients with *C. difficile* diarrhea and normal colonoscopy, as well as in patients with positive *C. difficile* cultures without diarrhea [19].

In one case report, longstanding carriage of *C. difficile* in a nursing home patient was associated with poor weight gain and low serum albumin despite aggressive nutritional support over many months, and these problems improved following weeks of treatment with oral vancomycin [4].

C. difficile infection among nursing home patients

The problem of *C. difficile* infection in long-term care facilities has received increasing attention in recent years following the first report of this infection among residents of a nursing home [3]. In this study, an initial point prevalence survey using stool culture found that 33 percent of patients on one ward of a 233-bed facility were colonized with *C. difficile* versus less than 10 percent of residents on the other wards. The affected ward housed the most debilitated residents in the facility including patients with severe pressure ulcers, closed head trauma, and endstage renal failure on dialysis. Serial surveys were carried out on this ward over the following 6 months, and the point prevalence rates remained high (33 to 47%) despite interventions including initial treatment with metronidazole or vancomycin of all patients with positive stool toxin assays, prescription of modified enteric isolation for all patients with positive stool toxin assays, and recommendations to the attending physicians to avoid specific classes of antibiotics thought to be most likely to predispose to *C. difficile* infection. This study raised the question of whether *C. difficile* infection was endemic in long-term care facilities.

In the years since this investigation, other studies and reports have shown that although *C. difficile* can be a significant problem among residents of nursing homes, this organism is not typically an endemic pathogen in long-term care facilities. Clusters of cases of *C. difficile* in nursing homes have been reported in Baltimore, Philadelphia, and Chicago [6], but true epidemic outbreaks of infection with rapid spread among the residents of a facility are probably rare. The epidemiology of *C. difficile* infection among nursing home patients might be best characterized as a scenario which includes occasional spread of infection from resident to resident, particularly to residents recently treated with antibiotics; frequent admission to the facility of individuals who are colonized with *C. difficile* [3, 17] ; and increased vigilance once initial cases are identified. The newly admitted patients may have had a previously treated bout of *C. difficile* diarrhea in the hospital, but most probably were simply colonized during the hospitalization, and may or may not develop symptomatic infection in the nursing home. When one or two cases of confirmed *C. difficile* diarrhea are identified in a facility, nursing and physician concern is heightened and more frequent testing for *C. difficile* is ordered.

Testing often occurs for all patients with any loose stools, and so individuals with *C. difficile* diarrhea which might be only minimal and self-limited, are identified as infected. Thus, nursing homes with a «problem» with *C. difficile* infection may be as likely to have increased identification of infected patients as an actual increased incidence and prevalence of disease.

Whether or not *C. difficile* is perceived as a problem in a long-term care facility, rigorous and rational attention to infection control in all nursing homes is necessary to minimize the morbidity from this infection (Table 1). Strict adherence to universal precautions for all residents, and regular in-service training for all staff members on the importance of handwashing will provide the foundation for infection control generally and for *C. difficile* infection specifically. In addition, rational prescription of antibiotics must be assured by the medical director and nursing quality control officer to minimize the risk that susceptible individuals are made more vulnerable to this potentially lethal infection. Although «sterilization» of the environment is an ideal, given the ubiquity of this organism and the hardiness of the spores it produces, a false sense of security will be engendered by those who place too great a reliance upon soap and germicidals alone for infection control. Special consideration should also be given to the communal use of electronic thermometers as spread of infection may be facilitated using these instruments [8].

Table 1. Prevention and control of *Clostridium difficile* outbreaks in nursing homes

- Provide regular inservice training on infection practices focused on handwashing
- Maintenance of high index of suspicion for *C. difficile* whenever a resident has diarrhea
- Use antibiotics rationally and avoid their use whenever possible
- Understand how *C. difficile* infection is diagnosed in your laboratory (toxin assay versus culture) and order appropriate tests
- Treat symptomatic patients promptly

Perhaps the most crucial component of an infection control system is maintaining a high index of suspicion for *C. difficile* disease. In a nursing home where many doctors may care for residents, the medical director and infection control nurse must often provide education to the physician staff about the approach to diagnosing and treating this infection. Since many new admissions and residents will have unrecognized carriage of this organism, a rational approach is recommended regarding the need of isolating those few patients who are known to be infected. A positive test for *C. difficile* should not be a basis for denying admission to a nursing home. Patients admitted to a facility who are asymptomatic and only have a positive test should be treated like an uninfected resident, except perhaps in a rare case of a demented patient who grossly soils the environment with feces. For patients who develop *C. difficile* diarrhea in the nursing home, treatment within the patient's room is appropriate, i.e., isolation in a private room is not necessary; soiled linens should be bagged at the bedside to minimize environmental contamination; and treatment in the room can be discontinued once the patient feels well, again unless fecal soiling of the environment is thought likely [5].

This rational approach to infection control in a long-term care facility differs from the stricter approach advocated for hospitalized patients by the Centers for Disease Control and Prevention, but has been adopted by the Maryland State Department of Health [16].

Treatment of *C. difficile* infection

Whether caring for patients in a hospital, nursing home, or office, prevention remains the key for minimizing the incidence of *C. difficile* disease. Although prevention involves attention to infection control as discussed above, avoiding the use of antibiotics is no doubt the most important single method for reducing the incidence of this infection generally. Since the establishment of *C. difficile* in the colon usually requires the disruption of the normal flora by antibiotics, the use of systemic antibiotics should be avoided whenever possible. For example, an elderly patient with an episode of vomiting with aspiration and a fever several hours later might be treated with aerosolized bronchodilators and chest percussion, one with an infected pressure sore and a low-grade fever might be treated with debridement and betadine wet-to-dry dressings, and another with a fever without obvious source who is otherwise clinically stable might be monitored and given extra fluids for 12 to 24 hours. If such patients become afebrile, as will often be the case, the need to prescribe antibiotics is obviated, and the risk of a frail patient developing *C. difficile* diarrhea from antibiotic treatment is avoided.

Whenever a patient who is being treated with antibiotics develops diarrhea, the antimicrobial agent should be discontinued if possible. Antibiotics are often arbitrarily ordered for 7 to 14 days for older individuals and for nursing home patients with dysuria, bronchitis or unexplained fevers, and when diarrhea develops after the fifth or sixth day of therapy serious consideration should be given to discontinuing antimicrobial treatment. Although theoretical concerns regarding incompletely treating an infection or predisposing to the emergence of resistant bacteria can be raised, discontinuation of the antibiotic will often be followed by cessation of diarrhea in 1 to 2 days with no return of the original symptoms for which antibiotics were prescribed. If the antibiotic regimen cannot be discontinued, additional therapy directed against *C. difficile* may be needed.

The initial approach to the treatment of an older patient with symptomatic *C. difficile* infection should be based on the severity of disease as defined by the volume of diarrheal stool losses [minimal (< 1 liter per day), moderate (1 to 2 liters per day) or severe (> 2 liters per day)]; the extent of fever [minimal (< 38°C), moderate (38 to 38.5°C) or severe (> 38.5°C)]; the extent of abdominal pain and distension [minimal (none), moderate (subjective complaints and/or identified by examination) or severe (significant pain and/or obvious distension with inspection)]; and the elevation of the white blood cell count [minimal (< 12,000/mm^3), moderate (12,000 to 25,000/mm^3) or severe (> 25,000/mm^3)].

Relying upon this simple approach to patient assessment, therapy for an elderly patient with *C. difficile* diarrhea can be prescribed. All patients should receive supportive therapy which includes careful attention to volume status and prescription of oral rehydration therapy (or intravenous fluids) to replace diarrheal fluid losses.

However, antiperistaltic drugs, e.g., diphenoxylate hydrochloride with atropine sulfate or loperamide, which can precipitate the development of pseudo-membranous colitis, should be avoided absolutely.

Patients with minimal symptoms can be managed simply with supportive therapy. Commercial oral rehydration therapy solutions include packets with pre-measured glucose and salts, e.g., Oral Rehydration Salts, Jianas Bros., Kansas City, MO; or pre-mixed solutions, e.g., Pedialyte, Ross Laboratories, Columbus, OH; Ricelyte, Mead Johnson, Evansville, IN; etc. As with any patient with mild gastroenteritis, those with minimally symptomatic *Clostridium difficile* diarrhea initially require 1 to 2 liters to replace diarrheal losses and intraluminal fluid once symptoms are identified, and an additional 1 to 2 liters each day to replace ongoing losses. In addition, the prescription of bismuth subsalicylate can be considered. This agent has antibacterial activity against *C. difficile in vitro* [11], and is pro-tective in an animal model of this infection [10]. There are no published clinical trials employing bismuth subsalicylate for individuals with *C. difficile* infection, but this medication can be safely given to elderly patients [Bennett R., unpublished results], and appears to be effective in reducing diarrheal symptoms in patients with *C. difficile* infection. Bismuth subsalicylate can be prescribed 30 ml every 2 hours up to 8 doses, and then 30 to 60 ml every 6 hours as needed for diarrhea over the following 3 to 4 days.

The two most frequently prescribed antibiotics to combat *C. difficile* infection in the United States are vancomycin and metronidazole. Because oral vancomycin is so expensive, most experts agree that metronidazole can be prescribed for patients who are clinically stable and have no severe symptoms as described above. For patients with moderate symptoms, metronidazole 500 mg every 6 hours for 7 to 10 days can be prescribed. In a comparison trial of metronidazole and vancomycin, both antibiotics were effective in alleviating symptoms of diarrhea over 3 to 5 days, and in preventing relapse which occurred in only 5 to 15 percent of subjects [20]. Although this study showed similar therapeutic efficacy, most specialists believe that oral vancomycin should always be used for patients with severe symptoms and for patients with confirmed pseudomembranous colitis. For patients with severe symptoms, vancomycin 500 mg every 6 hours for 10 to 14 days should be ordered. If symptoms improve quickly, the dose of vancomycin can be decreased to 250 mg after 4 to 5 days, and then to 125 mg after 10 days to minimize pharmacy costs.

Treating recurrent infection remains one of the challenges of dealing with *C. difficile* infection among the elderly. This is particularly true among the frailest nursing home patients who may develop recurrent disease over many months. Current approaches to treating these patients include careful attention to the overall nutritional state, and creative use of the available antimicrobials. For patients with intermittent but minimal diarrhea, e.g., loose stools 2 to 4 days each week, and persistently positive *C. difficile* laboratory tests, standing orders for bismuth subsalicylate as needed can be considered. For patients with deteriorating nutritional status or more troubling symptoms, prescription of tapering doses of vancomycin over many weeks may prove necessary. An exciting approach to treatment is the use of microbial replacement. The use of both *Saccharomyces boulardii* and *Lactobacillus GG* has been reported [13, 14], and ongoing investigations will confirm the efficacy of using *Saccharomyces boulardii* for treating patients with relapsing *C. difficile* disease.

Acknowledgement: This article was adapted from «*C. difficile* Infection in the Elderly» (1994) In: Powers D, Coe R, Morley JE (eds) Aging, Infection, and Immunity. Springer Publishing Company, New York, NY 10012, and is used by permission.

References

1. Aronsson B, Mollby R, Nord CE (1985) Antimicrobial agents and *Clostridium difficile* in acute enteric disease: epidemiological data from Sweden, 1980-1982. J Infect Dis 151: 476-481

2. Bartlett JG, Chang TW, Gurwith M, Gorbach SL, Onderdonk AB (1978) Antibiotic-associated pseudomembranous colitis due to toxin-producing clostridia. N Engl J Med 298: 531-534

3. Bender BS, Bennett RG, Laughon BE, Greenough WB, Gaydos C, Sears S, Forman M, Bartlett JG (1986) Is *Clostridium difficile* endemic in chronic-care facilities? Lancet ii: 11-13

4. Bennett RG, Greenough WB (1990) *C. difficile* diarrhea: a common and overlooked nursing home infection. Geriatrics 45: 77-87

5. Bennett RG, Greenough WB (1994) Diarrhea. In: Hazzard WR, Bierman EL, Blass JP, Ettinger WH, Halter JB (eds) Principles of Geriatric Medicine and Gerontology, 3rd edn. McGraw Hill, New York

6. Bennett RG, Laughon BE, Greenough WB, Bartlett JG (1989) *Clostridium difficile* in elderly patients (letter). Age Ageing 18: 354-355

7. Bobo L, Gaydos C, Bennett RG, Laughon BE, Bartlett JG (1987) Comparison of methods for typing *Clostridium difficile* (Abstract C-10) Annual Meeting, American Society of Microbiology

8. Brooks SE, Veal RO, Kramer M, Dore L, Schupf N, Adachi M (1992) Reduction in the incidence of *Clostridium difficile*-associated diarrhea in an acute care hospital and a skilled nursing facility following replacement of electronic thermometers with single-use disposables. Infect Control Hosp Epidemiol 13: 98-103

9. Brown E, Talbot GH, Axelrod R, Provencher M, Hoegg C (1990) Risk factors of *Clostridium difficile* toxin-associated diarrhea. Infect Control Hosp Epidemiol 11: 283-290

10. Chang TW, Dong MY, Gorbach SL (1990) Effect of bismuth subsalicylate on *Clostridium difficile* colitis in hamsters. Rev Inf Dis 12 suppl 1: S57-58

11. Cornick NA, Silva M, Corbach SL (1990) *In vitro* activity of bismuth subsalicylate. Rev Inf Dis 12 suppl 1: S9-10

12. George RH, Symonds JM, Dimock F, Brown JD, Arabi Y, Shinagawa N, Keighley MRB, Alexander-Williams J, Burdon DW (1978) Identification of *Clostridium difficile* as a cause of pseudomembranous colitis. Br Med J 1: 695

13. Gorbach SL, Chang TW, Goldin B (1987) Successful treatment of relapsing *Clostridium difficile* colitis with *Lactobacillus GG* (letter). Lancet ii: 1519

14. Kimmey MD, Elmer GW, Surawicz CM, McFarland LV (1990) Prevention of further recurrences of *Clostridium difficile* colitis with *Saccharomyces boulardii*. Dig Dis Sci 35: 897-901

15. Kuhl SJ, Tang YJ, Navarro L, Gumerlock PH, Silva Jr J (1993) Diagnosis and monitoring of *Clostridium difficile* infections with the polymerase chain reaction. Clin Infect Dis 16: S234-238

16. Maryland Department of Health and Mental Hygiene (October 1989) *Clostridium difficile* in long-term care facilities: recommendations for control and for admission of residents. Baltimore, Maryland

17. McFarland LV, Mulligan ME, Kwok RYY, Stamm WE (1989) Nosocomial acquisition of *Clostridium difficile* infection. N Engl J Med 320: 204-210

18. McFarland LV, Surawicz CM, Stamm WE (1990) Risk factors for *Clostridium difficile* carriage and *C. difficile*-associated diarrhea in a cohort of hospitalized patients. J Infect Dis 162: 678-684

19. Rybolt AH, Bennett RG, Laughon BE, Thomas DR, Greenough WB, Bartlett JG (1989) Protein-losing enteropathy associated with *Clostridium difficile* infection. Lancet i: 1353-1355

20. Teasley DG, Gerding DN, Olson MM, Peterson LR, Gebhard RL, Schwartz MJ, Lee JT Jr (1983) Prospective randomised trial of metronidazole versus vancomycin for *C. difficile*-associated diarrhoea and colitis. Lancet ii: 1043-1046

21. Thomas DR, Bennett RG, Laughon BE, Greenough WB, Bartlett JG (1990) Postantibiotic colonization with *Clostridium difficile* in nursing-home patients. J Am Geriatr Soc 38: 415-420

The clinical significance of *Clostridium difficile* infections in infants and children

Jean-Paul Buts

SUMMARY

In neonates, and in infants below the age of 6 months, the pathogenicity of *C. difficile* remains controversial. Cases of pseudomembranous colitis (PMC) due to *C. difficile* are rare before the age of two years, and some 60% of asymptomatic neonates excrete *C. difficile* in the stools because of natural nosocomial colonisation of the digestive tract during the first few days of life. After the age of one year, the asymptomatic carrier rate diminishes, with respect to the adult incidence, but remains at 5-10% until the age of two. Furthermore, many small infants who are asymptomatic excrete toxins A and B in their stools, in concentrations equal to or even higher than those seen in adults with PMC. The relative insensitivity of the infantile alimentary tract to toxins A and B may be the result of absent or immature intestinal receptors, or of the presence of maternal milk antibodies which neutralise the binding of toxin A to the receptor. Nonetheless, some toxin producing strains of *C. difficile* can be pathogenic in the young age group, as witnessed by the description of seven cases of fatal fulminant PMC in neonates, probably linked to direct invasion of the tissues by *C. difficile* with massive endotoxaemia. We should also mention one case of PMC with perforation of the colon in a neonate whose breast feeding mother was taking ciprofloxacin.

In infants older than 6 months, various types of chronic enteropathy, with or without colitis, due to *C. difficile* have been identified. They may be classified in two categories: without colitis (minor antibiotic diarrhea, chronic protracted or persistent diarrhea with malnutrition, states of overdistension with recurrent colic but without diarrhea) or with colitis (pseudomembranous colitis, fulminant colitis with toxic megacolon and perforation, and isolated *C. difficile* colitis without pseudomembranes).

In conclusion, the bowel problems seen in adults resulting from infection with *C. difficile*, such as pseudomembranous colitis, are rare during the first few years of life, although fulminating types of PMC which may be fatal, do occur in the neonate. In the young infant some toxin producing strains of *C. difficile* are responsible for chronic or relapsing enterocolitis, which respond to specific antimicrobial treatment.

Signification clinique des infections à *Clostridium difficile* chez les nourrissons et les enfants

RÉSUMÉ

Chez les nouveau-nés et les enfants de moins de 6 mois, la pathogénicité de *C. difficile* reste controversée. Les cas de colites pseudomembraneuses (CPM) dues à *C. difficile* sont rares avant l'âge de 2 ans, et plus de 60 % de nouveau-nés asymptomatiques excrètent *C. difficile* dans leurs selles, la colonisation naturelle du tube digestif s'effectuant dès les premiers jours de la vie par voie nosocomiale. Après l'âge d'un an, le portage asymptomatique diminue et l'incidence de *C. difficile* se rapproche de celle que l'on observe chez l'adulte, tout en restant d'environ 5 à 10 % jusqu'à l'âge de 2 ans. De plus, la plupart des tout-petits qui sont asymptomatiques, excrètent les toxines A et B dans les selles, en concentrations égales ou souvent supérieures à celles que l'on retrouve chez les adultes porteurs de CPM. L'insensibilité relative du tube digestif du nourrisson aux toxines A et B traduirait l'immaturité ou l'absence des récepteurs intestinaux, ou bien la présence dans le lait maternel d'anticorps capables de neutraliser la liaison de la toxine A à son récepteur. Quoi qu'il en soit, certaines souches de *C. difficile* toxinogènes peuvent être pathogènes dans les tranches d'âge les plus jeunes, comme en témoigne la description de sept cas fatals de CPM fulminantes chez des nouveau-nés, probablement en raison d'une invasion directe des tissus par *C. difficile* avec une endotoxinémie massive. Nous devons aussi signaler un cas de CPM avec perforation du côlon chez un nourrisson dont la mère, qui le nourrissait au sein, suivait un traitement par ciprofloxacine.

Chez les enfants de plus de 6 mois, différentes formes d'entéropathies chroniques, avec ou sans colite, ont été individualisées. Elles doivent être classées en deux catégories : sans colite (formes mineures de diarrhées associées aux antibiotiques, diarrhées chroniques rebelles ou persistantes avec malnutrition, hyperméatéorisme avec coliques récidivantes mais sans diarrhée.....); avec colite (entérocolite pseudo-membraneuse, colite fulminante avec mégacôlon toxique et perforation, colite avec présence de *C. difficile* mais sans pseudomembrane...).

En conclusion, les problèmes intestinaux vus chez l'adulte et résultant d'une infection à *C. difficile*, telle que la colite pseudomembraneuse, sont rares durant les toutes premières années de vie, bien que les formes fulminantes de CPM, qui peuvent être fatales, puissent survenir chez le nouveau-né. Chez le jeune enfant, certaines souches toxinogènes de *C. difficile* sont responsables d'entérocolites chroniques ou récidivantes, qui répondent à un traitement antimicrobien spécifique.

Die klinische Relevanz von *Clostridium difficile*-Infektionen bei Säuglingen und Kindern

ZUSAMMENFASSUNG

Bei Neugeborenen und Säuglingen bis zu 6 Monaten ist die Pathogenität von *C. difficile* nach wie vor umstritten. In den ersten zwei Lebensjahren sind Fälle von pseudomembranöser Kolitis (PMC) aufgrund von *C. difficile*-Infektionen selten, zumal rund 60 % der Neugeborenen aufgrund einer natürlichen nosokomialen Besiedelung ihres Verdauungstraktes während der ersten Lebenstage *C. difficile* mit dem Stuhl ausscheiden. Ab dem Alter von einem Jahr nimmt der Anteil symptomfreier Dauerausscheider im Vergleich zur Inzidenz bei Erwachsenen ab, liegt allerdings bis zum zweiten Lebensjahr noch bei 5-10 %. Darüber hinaus sind bei vielen symptomfreien Kleinkindern im Stuhl die Toxine A und B in ebenso hohen oder sogar höheren Konzentrationen als bei erwachsenen PMC-Patienten nachweisbar. Die relative Unempfindlichkeit des kindlichen Verdauungstraktes gegenüber den Toxinen A und B beruht möglicherweise auf noch fehlenden oder unreifen Darmrezeptoren oder auf der Aufnahme von Antikörpern mit der Muttermilch, welche die Rezeptorbindung von Toxin A verhindern. Dennoch können bestimmte toxinbildende *C. difficile*-Stämme auch bei Kleinkindern pathogen wirken, wie die berichteten sieben Fällen mit tödlich verlaufender fulminanter PMC bei Neugeborenen belegen; diese Fälle waren vermutlich auf eine direkte Gewebeinvasion von *C. difficile* mit massiver Endotoxämie zurückzuführen. Beachtenswert ist auch der Fall eines neugeborenen Kindes, bei dem es zu einer PMC mit Dickdarmperforation kam, da die stillende Mutter Ciprofloxacin einnahm.

Bei Kleinkindern über 6 Monate unterscheidet man mehrere *C. difficile*-bedingte chronische Darmerkrankungen mit und ohne Kolitis. Sie können in zwei Kategorien unterteilt werden: ohne Kolitis (geringgradige Antibiotika-assoziierte Diarrhö, chronische oder rezidivierende Diarrhö mit Fehlernährung, Fälle von Kolonerweiterung mit rezidivierenden Koliken, jedoch ohne Diarrhö) oder mit Kolitis (pseudomembranöse Kolitis, fulminante Kolitis mit toxischem Megakolon und Perforation sowie isolierte *C. difficile*-Kolitis ohne Pseudomembranen).

Zusammenfassend ist zu sagen, daß die Darmbeschwerden, die bei Erwachsenen infolge einer *C. difficile*-Infektion auftreten, beispielsweise die pseudomembranöse Kolitis, in den ersten Lebensjahren selten anzutreffen sind, wobei allerdings bei Neugeborenen fulminante Verläufe einer PMC mit letalem Ausgang auftreten können. Bei Kleinkindern sind bestimmte toxin-bildende *C. difficile*-Stämme für chronische oder rezidivierende Enterokolitiden verantwortlich, die auf eine spezifische Antibiotikabehandlung ansprechen.

Clostridium difficile infections in neonates and small children

Although *C. difficile* was described in 1935 by Hall and O'Toole, who isolated this strictly anaerobic organism from the stools of neonates [21] it was not until 1978 that its role as the causative agent in pseudomembranous colitis was identified [4, 26]. Thanks to the development of specific culture techniques for *C. difficile*, and the identification of toxin B in the hospital environment, it quickly became clear, towards the end of the 1970's, that this organism was responsible not only for a wide spectrum of digestive diseases and for the complications of antibiotic treatment, but also that the prevalence of contamination by *C. difficile* was particularly high in the hospital setting [3]. Colonization with *C. difficile* results from proliferation due to disturbance of the gut flora (mainly by antibiotics) either of nosocomial exogenous strains or of previously repressed endogenous ones.

Most of the available studies on the prevalence of *C. difficile* in the neonate and young infants have shown that neonates whose digestive tract flora is immature are naturally colonized by the organism during the first few days of life and that the asymptomatic carrier rate reaches 60% [7, 24]. Moreover, it has been shown in the hamster that neonate's stools stimulate *in vitro C. difficile* growth [6, 37]. The carrier rate remains high at 10-20% up to the age of two years, and then falls progressively until the adult value of below 5% is reached [14]. Most infections during the first week are nosocomial [29] as has been demonstrated in neonatal units by serotyping the strains [15, 16]. Spores can survive for months because of their resistance to oxygen, drying and conventional disinfectants [28, 31]. Although during the first two weeks of life the colonization rate is the same in breast fed babies as in those on artificial formula feeding, later the rate is much lower in the former group, and the difference becomes much more marked when breast feeding is prolonged and solid supplements are excluded [13, 40].

Many small infants who are asymptomatic but are colonized by *C. difficile*, excrete toxins A and B in their stools, in concentrations equal to or even higher than those seen in adults with pseudomembranous colitis [17, 35]. Indeed, at present, no correlation has been established between the excretion of toxin B and clinical manifestations of digestive disease [25, 27, 44]. The high incidence of asymptomatic carriers in neonates, in the absence of any signs of digestive disease and with no correlation with toxin production, reflects the rarity of pseudo-membranous colitis before the age of two years. These findings imply that considerable reservations must be made in regard to the pathogenicity of *C. difficile* in this age group, and that the detection of toxins A and B in the stools needs to be interpreted with caution [46].

Several hypotheses have been advanced to explain the high asymptomatic carrier rate in the neonatal period.

(1) Some authors [42] have suggested that the relative resistance of the neonatal digestive tract to the toxins of *C. difficile* may be the result of the absence or immaturity of the specific receptors. The secretory and inflammatory effects of *C. difficile* seem to be produced solely by toxin A, a protein of 308 kD with powerful enterotoxic and cytotoxic actions. In the newborn rabbit, Eglow et al [19] have shown that the relative resistance of the ileum to toxin A is associated with a low concentration of receptors. In contrast, Rolfe et al [36] have extracted from the brush border of the new-born hamster a functional and specific receptor for the toxin A of *C. difficile*. The receptor is a glycoprotein linked to protein G and contains a N-acetylglucosamine group and galactose, the molecular weight of which is around 163 kD [33, 47].

(2) Other authors have proposed a neutralising effect of the maternal antibodies on the toxin/receptor complex. In particular, Rolfe et al [38] have recently shown that the immunoglobulins (> 90% IgA) and a non-globulin component of the maternal milk, exert a neutralising effect on the binding of *C. difficile* toxin A to the receptor.

(3) Finally, it is always possible that the inhibited inflammatory response to the toxins in the colon of the neonate may simply reflect the immaturity of the immune system.

In contrast, a number of experimental and clinical observations indicate that in some rare cases the *C. difficile* enterotoxins may be very virulent in the new-born, as witnessed by the description of seven cases of fatal fulminant pseudo-membranous colitis, probably linked to direct invasion of the tissues by *C. difficile* with massive endotoxemia [34].

There is also one patient with pseudomembranous colitis complicated by colonic perforation whose lactating mother was taking ciprofloxacin [23]. High concentrations of the drug were detected in the milk [10, 23]. The direct toxicity of toxins A and B for organs outside the gut is seen in the Rhesus monkey, in which the injection of a few micrograms of toxins produces sudden death due to cerebral failure [1]. Thus clinical and experimental studies are essential for defining the pathogenic mechanism of *C. difficile* and its toxins, which appear remarkably different in the neonate and the adult.

C. difficile infections in infants older than six months and children

In infants older than 6 months, various types of chronic enteropathy, with or without colitis, due to *C. difficile* have been identified. They may be classified as follows:

• **Without colitis**
  - minor antibiotic diarrhea
  - chronic or recurrent diarrhea with malnutrition
  - states of bowell overdistension with recurrent episodes of colics but without diarrhea
• **With colitis**
  - pseudomembranous colitis
  - fulminant colitis with toxic megacolon and perforation
  - isolated *C. difficile* colitis without pseudomembranes

In the infant and young child, the pathogenic role of *C. difficile* seems to be implicated in some cases of necrotizing enterocolitis [11, 22], in enterocolitis occurring before or after operations for Hirschsprung's disease [2], in the intestinal complications of the hemolytic/uremic syndrome and in some cases of recurrent ulcerative colitis [30, 43].

Several studies have suggested a role for enterotoxic strains of *C. difficile* in the infant sudden death syndrome (SIDS), but this remains controversial. In 1980, Scopes et al [39] reported a case of sudden infant death syndrome in a child suffering from *C. difficile* pseudomembranous colitis, where the colitis was found by chance at autopsy. Cooperstock et al [12] reported high levels of the toxins of *C. difficile* in the stools of two patients with SIDS.

In a study which included 123 sudden infant deaths, Murrell et al [32] recorded high levels of *C. difficile* toxins in the stools and the serum of these patients as compared with 52 controls (Table 1). Although there were significant differences between the SIDS and the control groups, the real importance of these differences remains obscure, and it would be unwise at the present time to put too much importance on the role of *C. difficile* in the etiology of this lethal syndrome.

Table 1. Prevalence of *C. difficile* toxin in the serum in sudden infant death syndrome (SIDS) and in control subjects

Bacteria	Patients	
	SIDS	Controls
C. perfringens	54/119 (45%)	10/51 (19%)
C. difficile	33/119 (27%)	8/54 (14%)
C. botulinum	6/120 (5%)	0/53 (0%)

The clinical picture of pseudomembranous colitis in children has been known for over 20 years, well before the identification of the pathogenic role of *C. difficile* in 1978 [9, 20]. The clinical features are much the same as those in the adult and comprise: profuse bloody diarrhea, intense abdominal pain, hyperthermia, vomiting (rarely), edema and ascites resulting from protein losing enteropathy, and a neutrophil leucocytosis. The most frequent age range is from 5 to 17 years. The great majority of cases (over 90%) of pseudomembranous colitis are due to an enterotoxin producing strain of *C. difficile* that appears during the course of antibiotic treatment, especially with penicillins and cephalosporins. Lincosamines are at the greatest risk, but are rarely prescribed in children [5].

In a recent clinical study [8] we described a series of 19 children aged from two months to 11 years (median 8 months) all of whom presented with bowel problems related to *C. difficile*. The series comprised 8 children with chronic protracted and persistent diarrhea requiring parenteral feeding, 4 with distension, recurrent episodes of colics and vomiting, and 7 with a mixed picture. All were carriers of pathogenic strains of *C. difficile* identified by serotype and excreted high amounts of toxins A and B. A prospective therapeutic trial of oral *Saccharomyces boulardii* over 15 days resulted in a rapid resolution of symptoms in 95% of cases, and eradication of the toxin in 95% with elimination of *C. difficile* from the stools in 85%.

Conclusions

• The bowel problems seen in adults resulting from infection with *C. difficile*, such as pseudomembranous colitis are rare during the first few years of life.

• Fulminating types of pseudomembranous colitis with perforation of the colon, do occur in the neonate. The physiopathology is obscure, but may be related to tissue invasion with resulting massive endotoxemia.

• In the young infant some toxin producing strains of *C. difficile* are responsible for chronic or relapsing enterocolitis, which respond to specific antimicrobial treatment [8, 41].

References

1. Arnon SS, Mills DC, Day PA, Henrickson RV, Sullivan NM, Wiskins TD (1984) Rapid death of infant Rhesus monkeys injected with *Clostridium difficile* toxin A and B: physiologic and pathologic basis. J Pediatr 104: 34-40
2. Bagwell CE, Langhan MR JR, Mahaffey SM, Talbert JL, Shandling B (1992) Pseudomembranous colitis following resection for Hirschsprung's disease. J Pediatr Surg 27: 1261-1264
3. Bartlett JG (1988) Introduction. In: Rolfe RD, Finegold SM (eds) *Clostridium difficile* Its role in intestinal disease. Academic Press, San Diego, pp 1-13
4. Bartlett JG, Chang TW, Gurwith M, Gorbech SL, Onderdonk AB (1978) Antibiotic-associated pseudomembranous colitis due to toxin-producing clostridia. N Engl J Med 298: 531-534
5. Beesley J. Eastman EJ, Jackson RH, Nelson R (1981) Clindamycin associated pseudo-membranous colitis. Acta Pediatr Scand 70: 129-130
6. Borriello SP, Barclay FE (1986) An *in vitro* model of colonisation resistance to *Clostridium difficile* infection. J Med Microbiol 21: 299-309
7. Bretfle RP, Wallace E (1982) *Clostridium difficile* from stools of normal children. Lancet i: 1193
8. Buts JP, Corthier G. Delmée M (1993) *Saccharomyces boulardii* for *Clostridium difficile*-associated enteropathies in infants. J Pediatr Gastroenterol Nutr 16: 419-425
9. Buts JP, Weber AM, Roy CC, Morin C (1977) Pseudomembranous enterocolitis in childhood. Gastroenterology 73: 823-827

10. Cain DB, O'Connor ME (1990) Pseudomembranous colitis associated with ciprofloxacin. Lancet 336: 1509-1510

11. Cashore WJ, Peter G. Lauerman M, Stonestreet BS, Oh W (1981) Clostridial colonization and clostridial toxin in neonatal necrotizing enterocolitis. J Pediatr 98: 308-311

12. Cooperstock M, Riegle L, Fabacker D, Woodruff CW (1982) *Clostridium difficile* in formula-fed infants and sudden infant death syndrome. Pediatrics 70: 91-95

13. Cooperstock M, Riegle L, Woodruff CW, Onderdonk A (1983) Influence of age, sex and diet upon asymptomatic colonization of infants with *Clostridium difficile*. J Clin Microbiol 17: 830-833

14. Delmée M, Buts JP (1993) *Clostridium difficile*-associated diarrhoea in children. In: Buts JP, Sokal EM (eds) Management of Digestive and Liver Disorders in Infants and Children. Elsevier, Amsterdam, pp 371-379

15. Delmée M, Homel M, Wauters G (1985) Serogrouping of *Clostridium difficile* strains by slide agglutination. J Clin Microbiol 59: 1192-1195

16. Delmée M, Verellen G. Avesani V, François G (1988) *Clostridium difficile* in neonates: serogrouping and epidemiology. Eur J Pediatr 147: 36-40

17. Donta ST, Meyers MG (1982) *Clostridium difficile* toxin in asymptomatic neonates. J Pediatr 100: 431-434

18. Dove CH, Wang SZ, Price SB, Phelps CJ, Lyerly DM, Wilkins TD, Johnson JL (1990) Molecular characterization of the *Clostridium difficile* toxin A gene. Infect Immun 58: 480-488

19. Eglow R, Pothoulakis C, Itzkowitz, Israël EJ, O'Keane CJ, Gong D, Gao N, Xu L, Walker A, LaMont JT (1992) Diminished *Clostridium difficile* toxin A sensitivity in newborn rabbit ileum is associated with decreased toxin A receptor. J Clin Invest 90: 822-829

20. Fee HJ, Kearny JL, Ament ME (1975) Fatal outcome in a child with pseudomembranous colitis. J Pediatr Surg 10: 959-963

21. Hall J. O'Toole E (1935) Intestinal flora in newborn infants with description of a new pathogenic organism, *Bacillus difficilis*. Am J Dis Child 49: 390-402.

22. Han VKM, Sayed H. Chance GW, Brabyn DG, Shaheed WA (1983) An outbreak of *Clostridium difficile* necrotizing enterocolitis: a case for oral vancomycin therapy? Pediatrics 71: 935-941

23. Harmon T, Burkhart G. Applebaum H (1992) Perforated pseudomembranous colitis in the breast-fed infant. J Pediatr Surg 27: 744-746

24. Holst E. Helin I. Mardh PA (1981) Recovery of *Clostridium difficile* from children. Scan J Infect Dis 13: 41-45

25. Kotloff KL, Wade JC, Morris JG (1988) Lack of association between *Clostridium difficile* toxin and diarrhoea in infants. Pediatr Infect Dis J 7: 662-663

26. Larson HE, Honour P, Price AB (1978) *Clostridium difficile* and the etiology of pseudomembranous colitis. Lancet i: 1063-1066

27. Libby JM, Donta ST, Wilkins TD (1983) *Clostridium difficile* toxin A in infants. J Infect Dis 148: 606

28. Malamov-Ladas H, O'Farrell S, Nasj JQ, Tabaqchali S (1983) Isolation of *Clostridium difficile* from patients and the environment of hospitals. J Clin Pathol 36: 88-92

29. McFarland LV, Mulligan ME, Kwok RYY, Stamm WE (1989) Nosocomial acquisition of *Clostridium difficile* infection. N Engl J Med 320: 204-210

30. Meyers S. Mayer L, Bottone E, et al (1981) Occurrence of *Clostridium difficile* toxin during the inflammatory bowel disease. Gastroenterology 80: 693-696

31. Mulligan ME, Rolfe RD, Finegold SM, George WL (1979) Contamination of a hospital environment by *Clostridium difficile*. Curr Microbiol 3: 173-175

32. Murrell WG (1993) Enterotoxigenic bacteria in the sudden infant death syndrome. J Med Microbiol 39: 114-127

33. Pothoulakis C, LaMont JT, Eglow R (1991) Characterization of rabbit ileal receptors for *Clostridium difficile* toxin A. Evidence for a receptor-coupled G protein. J Clin Invest 88: 119-125

34. Qualmon SJ, Petric M, Karmali MA, Smith CR, Hamilton SR (1990) *Clostridium difficile* invasion and toxin circulation in fatal pediatric pseudomembranous colitis. Am J Clin Biol Pathol 94: 410-416

35. Richardson SA, Alcock PA, Gray J (1983) *Clostridium difficile* and its toxin in healthy neonates. Br Med J 287: 878

36. Rolfe RD (1993) Purification of a functional receptor for *Clostridium difficile* toxin A from intestinal brush border membranes of infant hamsters. Clin Infect Dis 16 suppl 4: S219-227

37. Rolfe RD, Iaconis JP (1983) Intestinal colonization of infant hamsters with *Clostridium difficile*. Infect Immun 42: 480-486

38. Rolfe RD, Song W (1995) Immunoglobulin and non-immunoglobulin components of human milk inhibit *Clostridium difficile* toxin A receptor binding. J Med Microbiol 42: 10-19

39. Scopes JW, Smith MF, Beack RC (1980) Pseudomembranous colitis and sudden infant death. Lancet i: 1144

40. Stark PL, Lee A, Parsonage BD (1982) Colonisation of the large bowel by *Clostridium difficile* in healthy infants: quantitative study. Infect Immun 35: 895-899

41. Stuphen JL, Grand RJ, Flores A, Chang TW, Bartle HJG (1983) Chronic diarrhoea associated with *Clostridium difficile* in children. Am J Dis Child 137: 275-278

42. Triadafilopoulos G, LaMont JT (1993) Pseudomembranous colitis. In: Walker, Durie, Hamilton, Walker-Smith, Watkins (eds) Pediatric Gastrointestinal Disease (vol 1). BC Decker Inc, Philadelphia, pp 619-629

43. Trnka YM, LaMont JT (1981) Association of *Clostridium difficile* toxin with symptomatic relapse of chronic inflammatory bowel disease. Gastroenterology 80: 693-696

44. Ushijima H, Shinozaki T, Fujii R (1985) Detection of *Clostridium difficile* exterotoxin in neonates by latex agglutination. Arch Dis Child 60: 252-271

45. Von Eichel-Streiber C, Sauerbom M (1990) *Clostridium difficile* toxin A carriers a C-terminal repetitive structure homologous to the carbohydrate binding region of *Streptoccocus* glycosyltransferases. Gene 96: 107-113

46. Welch DF, Marks MJ (1982) Is *Clostridium difficile* pathogenic in infants? J Pediatr 100: 393-395

47. Wilkins TD, Tucker KD (1989) *Clostridium difficile* toxin A (enterotoxin) uses GALa1-3GALß1-4GLCNAC as a functional receptor. Microecology and therapy 19: 225-227

Nocosomial acquisition and risk factors for *Clostridium difficile* disease

Lynne V. McFarland

SUMMARY

Clostridium difficile is the most frequent known cause of infectious diarrhea which is acquired in the hospital by adult patients. Outbreaks of *C. difficile* are likely to occur in the hospital because patients who are normally resistant to *C. difficile* colonization become susceptible when they are exposed to antibiotics or medications which disrupt their normal colonic flora. In addition, the newly susceptible patients are hospitalized in an environment where sources of *C. difficile* are clustered (infected patients and contaminated environmental sites). The detection of *C. difficile* outbreaks is often delayed because it is not suspected due to the fact that diarrhea (the most common symptom of *C. difficile*) is frequent among hospitalized patients due to a wide variety of etiologies. Once the outbreak starts, *C. difficile* may be rapidly spread and become disseminated throughout the hospital environment where spores may persist for months. The number of reported outbreaks increased from less than 10 per year in the early 1980's to 10-20 per year from 1986-1993 and by 1994, over 30 outbreaks were reported. Not only has the frequency of nosocomial outbreaks increased, but *C. difficile* has become a problem in most European countries in addition to a known problem in the United States and the United Kingdom. The frequency of *C. difficile* has also increased in the pediatric, immunocompromised and extended care patient populations. The impact of these outbreaks is reflected by longer lengths-of-stays in hospitals (mean of 8 days), increased morbidity (4-5 times), an increased risk of mortality (2-3 times) and increased medical costs ($2000-5000 per episode). The routes of transmission in the hospital include exogenous transmission by contaminated environmental surfaces, dust containing spores, shared instrumentation, carriage on hospital personnel hand surfaces and infected roommates. Patients may also harbor spores of *C. difficile* and the endogenous carriage becomes symptomatic only after nosocomial exposures disrupt normal colonic flora, making the patient susceptible to *C. difficile* overgrowth. Risk factors include most types of medium to broad spectrum antibiotics, medications or procedures which disrupt normal bowel flora and a variety of host factors. The methods of controlling a nosocomial outbreak depend upon breaking the routes of transmission by attempting to identify the common source, through environmental decontamination, adherence to rigorous infection control practices and prompt diagnosis and treatment of infected patients. Prevention of future nosocomial outbreaks involves routine surveillance for *C. difficile*, education programs stressing the importance of handwashing or vinyl glove use and continued efforts to eliminate the persistent spores in the environment.

Contamination nosocomiale et facteurs de risque de l'infection à *Clostridium difficile*

RÉSUMÉ

Clostridium difficile est la plus fréquente des causes identifiées de diarrhée infectieuse acquise à l'hôpital par des patients adultes. En effet, les patients qui sont normalement résistants à la colonisation par *C. difficile* deviennent plus fragiles lorsqu'ils sont exposés à des antibiotiques ou à des médicaments qui perturbent l'équilibre de la flore intestinale normale. De plus, les patients fragilisés par une hospitalisation récente arrivent dans un environnement où les sources de *C. difficile* sont regroupées (patients infectés et contamination de l'environnement). La détection des épidémies de *C. difficile* est souvent différée car elles ne sont pas suspectées. En effet, la diarrhée, qui est le symptôme le plus fréquent de l'infection à *C. difficile*, est fréquente parmi les patients hospitalisés et peut relever de multiples causes.

Une fois que l'épidémie a commencé, *C. difficile* peut se propager rapidement, et disséminer à tout l'environnement hospitalier où les spores peuvent persister des mois. Le nombre d'épidémies rapportées est passé de moins de 10 par an dans les années 80, à 10 à 20 par an pour les années allant de 1986 à 1993. En 1994, plus de 30 épidémies ont été rapportées. La fréquence des infections nosocomiales a non seulement augmenté, mais *C. difficile* est devenu un problème de santé dans la plupart des pays européens, à l'instar des États Unis et du Royaume Uni, où les difficultés posées par *C. difficile* étaient déjà bien connues. La fréquence de *C. difficile* a aussi augmenté en pédiatrie, chez les sujets immunodéprimés ou chez les patients en long séjour. L'impact de ces épidémies se traduit par différents éléments :
- l'allongement de la durée du séjour dans les hôpitaux (8 jours en moyenne);
- l'augmentation de la morbidité (multipliée par 4 ou 5);
- l'augmentation du risque de mortalité (multiplié par 2 ou 3);
- l'augmentation du coût médical (2000 à 5000 $ US par épisode).

À l'hôpital, les voies de transmission incluent la transmission exogène par les surfaces environnementales contaminées, les spores contenues dans la poussière, l'appareillage médical réutilisable, le portage sur les mains du personnel hospitalier, les chambres infectées. Les patients peuvent aussi héberger les spores de *C. difficile*, et le portage endogène devient symptomatique si des contaminations nosocomiales perturbent la flore intestinale normale, exposant le patient à la croissance de *C. difficile*. Les facteurs de risque incluent la plupart des antibiotiques à spectre moyen ou large, les médicaments et gestes thérapeutiques qui perturbent la flore intestinale normale, et une grande variété de facteurs hospitaliers. Les méthodes de contrôle des épidémies nosocomiales reposent sur différents points : interruption des voies de transmission en essayant d'identifier leur source, décontamination de l'environnement, pratique rigoureuse du contrôle des infections, diagnostic rapide et respect du traitement des patients infectés. La prévention des futures épidémies nosocomiales impliquent la détection en routine de *C. difficile*, des programmes d'éducation insistant sur l'importance du lavage des mains et de l'usage de gants en vinyle, et des efforts continus pour éliminer les spores résistantes de l'environnement.

Nosokomiale Infektion und Risikofaktoren
für eine *Clostridium difficile*-Erkrankung

ZUSAMMENFASSUNG

C. difficile ist der gängigste bekannte Auslöser infektiöser Diarrhöen, mit dem sich erwachsene Patienten im Krankenhaus infizieren. *C. difficile*-Erkrankungen treten überwiegend im Klinikbereich auf, weil Patienten, die normalerweise resistent gegen eine Besiedelung mit *C. difficile* sind, bei Einnahme von Antibiotika oder anderen Medikamenten, die ihre normale Darmflora schädigen, anfällig für eine Infektion werden. Darüber hinaus sind die anfällig gewordenen Patienten im Krankenhaus multiplen Ansteckungsmöglichkeiten mit *C. difficile* ausgesetzt (infizierte Patienten und kontaminierte Umgebung). Der Nachweis von *C. difficile*-Erkrankungen verzögert sich oft, weil keine Verdachtsmomente vorliegen, denn Diarrhöen (häufigstes Symptom für *C. difficile*) sind bei stationären Patienten aufgrund vielfältiger Ätiologien ohnehin häufig. Nach dem Ausbrechen der Krankheit breitet sich *C. difficile* oft schnell aus und wird durch die gesamte Krankenhausumgebung gestreut, wo sich die Sporen monatelang halten können. Die Zahl der berichteten Epidemien stieg von unter 10 pro Jahr in den frühen 80er Jahren auf 10-20 pro Jahr zwischen 1986 und 1993 an, 1994 wurden sogar über 30 solche Ausbrüche gemeldet. Doch nicht nur die Häufigkeit nosokomialer Epidemien ist gestiegen, sondern *C. difficile* ist darüber hinaus in den meisten europäischen Ländern zu einem ebensolchen Problem geworden, wie es in den USA und Großbritannien schon seit längerem bekannt ist. Auch bei pädiatrischen, immungeschwächten und schwer pflegebedürftigen Patienten nimmt die Häufigkeit von *C. difficile*-Erkrankungen seit einiger Zeit zu. Die Auswirkungen solcher Krankheitsausbrüche lassen sich an den längeren stationären Aufenthalten (im Mittel 8 Tage), der erhöhten Morbidität (auf das 4-5fache), einem erhöhten Mortalitätsrisiko (auf das 2-3fache) und gestiegenen Behandlungskosten ($2.000-5.000 pro Episode) ablesen. Zu den Übertragungswegen im Krankenhaus gehören die exogene Ansteckung durch kontaminierte Oberflächen, sporenbefrachteten Staub, für mehrere Patienten benutzte Instrumente, kontaminierte Hände des Pflegepersonals und infizierte Bettnachbarn. Es kommt auch vor, daß Patienten Träger von *C. difficile*-Sporen sind und es erst durch den nosokomialen Kontakt zu einer Störung der Darmflora kommt, die den Patienten für ein überschießendes Wachstum von *C. difficile* anfällig macht und Symptome hervorruft. Zu den Risikofaktoren gehören die meisten Breitbandantibiotika, Medikamente oder Verfahren, bei denen die gesunde Darmflora geschädigt wird, sowie verschiedene den Wirtsorganismus betreffende Faktoren. Zur Eindämmung nosokomialer Infektionen müssen die Übertragungswege unterbrochen werden, und zwar durch Bemühungen, die gemeinsame Ursache zu identifizieren, durch Desinfektion der Umgebung, Befolgung strenger Methoden zur Infektionsverhütung und umgehende Diagnose und Behandlung infizierter Patienten. Um nosokomiale Epidemien künftig verhüten zu können, muß eine routinemäßige Prüfung auf *C. difficile* ebenso erfolgen wie Personalschulungen, bei denen die Bedeutung des Händewaschens oder des Tragens von Vinylhandschuhen vermittelt werden; außerdem muß ständig versucht werden, etwa vorhandene Sporen aus der Umgebung zu eliminieren.

Introduction

Clostridium difficile is the most frequent known cause of nosocomial outbreaks of diarrhea in adult hospitalized patients. Nosocomial outbreaks of *Clostridium difficile* began to be reported in 1980 (Table 1). Since 1980, interest in the problem of nosocomial *Clostridium difficile* infections has increased, as reflected by the steady increase in the number of citations regarding nosocomial *Clostridium difficile* - associated disease (CDAD) (Fig. 1).

Clostridium difficile outbreaks are not just confined to hospitals in the USA and United Kingdom hospitals, but are now reported in countries distributed around the world. From 1980 to 1993, an average of 4 hospital outbreaks were reported per year; but in 1994, the number of outbreaks dramatically increased to 12. Once an outbreak occurs, *Clostridium difficile* may persist in the hospital environment despite active infection control measures and this persistence may give rise to further outbreaks. Several studies have shown chronic, endemic persistence of *Clostridium difficile* up to ten years after an initial outbreak [6, 10, 44, 47].

Studies of *Clostridium difficile* outbreaks reported in the literature may be difficult to compare, due to different case definitions. Most well-designed studies have defined nosocomial as: new onset of diarrhea which was not present within 72 hours of admission; or if the patient was symptomatic on enrollment, the patient had a recent stay (within 30 days) at a hospital [19, 35, 37]. *Clostridium difficile*-associated disease has been defined through the fulfillment of three criteria: (1) diarrhea of three or more watery-loose stools per day for more than two days (or 6-8 loose-watery stools within 48 hours); (2) diarrhea in association with at least one positive *Clostridium difficile* assay (culture, toxin A or toxin B) and (3) the

Table 1. Selected nosocomial outbreaks of *Clostridium difficile*-associated disease (CDAD) in hospitalized patients

Year	Reference	
1980	[8]	Hospital outbreak reported.
1982	[62]	Phenotypically similar strains in 12/16 surgery patients.
1983	[31]	Antibiogram patterns show cross-infection
1984	[57]	Outbreak in 2 hospitals (radio PAGE Group «X»).
1984	[50]	Orthopedic ward outbreak (Immunochemical fingerprinting)
1986	[15]	Orthopedic surgical unit (Serogroup C)
1986	[19]	Outbreak in surgical unit
1986	[2]	Nursing home outbreak
1988	[22]	Medical ward (2 bacteriophage-bacteriocin strains)
1989	[38]	Orthopedic ward (SDS Page strain E)
1989	[18]	ICU (time-space cluster)
1989	[37]	General medicine ward (Immunoblot type 1)
1990	[21]	Gerding's 1986 outbreak (REA type B/B2)
1991	[56]	Surgical ward outbreak (Serotype C)
1992	[11]	Geriatric ward (Pyrolysis mass spectrometry)
1993	[30]	Elderly patients (PMS & antibiograms)
1994	[10, 23, 41, 44, 45]	Five hospital outbreaks reported
1994	[1]	Outbreak in AIDS patients (1 RAPD pattern)

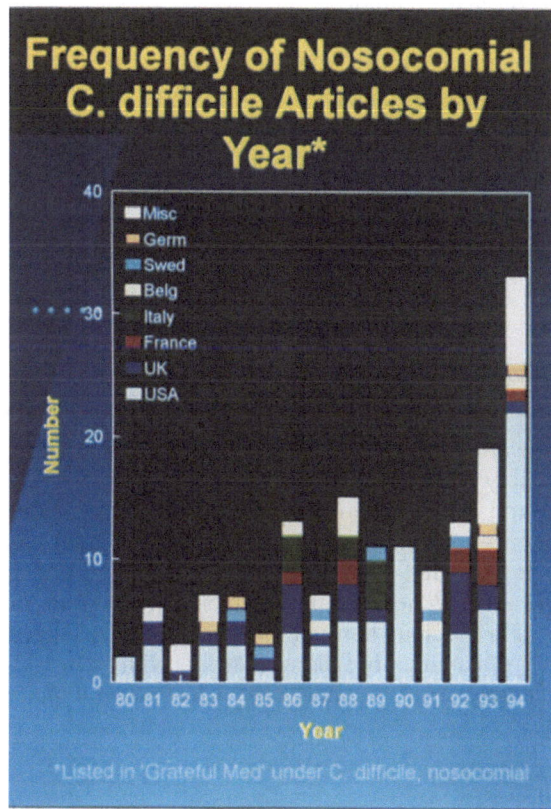

Fig. 1.
« Frequency of nosocomial *Clostridium difficile* articles by year » secular trends show *C. difficile* outbreaks are increasing both by frequency and country of origin

exclusion of other etiologies (medications, chronic intestinal conditions, etc.) of diarrhea [19, 28, 37, 47]. Outbreaks have been defined as a sudden increase in frequency of CDAD (greater than the endemic rate, if present) due to one strain type and isolated from patients which have a temporal-spatial relationship.

Impact of
nosocomial CDAD outbreaks

The impact of nosocomial outbreaks is reflected by longer lengths-of-stay, additional morbidity and mortality and higher hospital costs associated with CDAD. In adult inpatients, infection with *Clostridium difficile* has been shown to increase lengths-of-stay by an average of 8 days and in elderly patients, stays increased by an average of 36 additional days [16, 37]. The longer hospital stays incur additional costs, may expose patients to other nosocomial pathogens and pose an additional threat for *Clostridium difficile* transmission to other susceptible patients [35]. Patients with CDAD are found to have mortality rates 2-3 times the rate of similar patients matched on age, diagnosis, and severity of underlying diseases [16, 19].

The range of mortality rates in patients with CDAD has been from 2% to 21% in adult inpatients and 38% to 83% in geriatric patients [2, 16, 33, 59].

The costs of nosocomial CDAD include hospital per diem charges, microbiologic laboratory assays, endoscopic examinations and antibiotic treatments [26, 47]. A study in 1991 estimated the overall charges for an admitting diagnosis of CDAD averaged $5,000.00 per admission, whereas acquired cases resulted in additional hospital costs of $2,000.00 [26].

Transmission of *Clostridium difficile*

The success of *Clostridium difficile* as a nosocomial pathogen is attributable to the ease of transmission along a variety of routes, the persistence of *Clostridium difficile* spores in the hospital and the availability of a susceptible patient population. The sources of *Clostridium difficile* may be either endogenous (as normal colonic flora) or exogenous (acquired from outside the host), as shown in Fig. 2. A person may carry *Clostridium difficile* as part of their normal flora, or may have become colonized during a previous hospital stay and been discharged without developing disease. Studies of community sources (patients with no history of previous hospital stays) indicate only 2-7% of adults carry *Clostridium difficile*

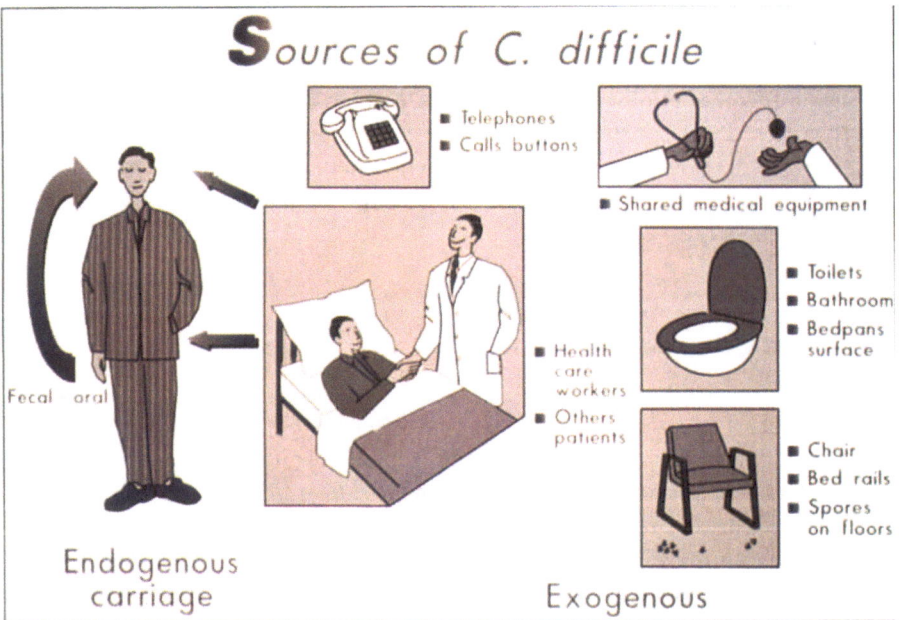

Fig. 2. Sources of *Clostridium difficile*

[12, 37, 54]. In patients with recent hospitalization (less than 30 days) who were cultured upon re-admission, 17-21% have been found to be harboring *Clostridium difficile* [12, 37, 54]. However for the majority of patients, *Clostridium difficile* is acquired from exogenous sources. These exogenous sources may include a wide variety of contaminated environmental sites (usually bathrooms and surfaces near the patient's beds), hands of hospital personnel (nurses, physicians, housekeepers, etc.), and even other patients [17, 22, 31, 34, 42, 55]. The main culprit for the transmission of nosocomial CDAD is the spore of *Clostridium difficile*. This spore is shed from asymptomatic carriers (often an unsuspected source of outbreaks) and from patients with diarrhea. The spores have been shown to remain viable for up to 5 months on environmental surfaces and are resistant to most commonly used hospital cleaning agents [17, 25, 43]. One study sampled 216 environmental sites on a general medicine ward and found 62 (29%) were positive for *Clostridium difficile* [37].

The frequency of environmental contamination was found to be associated with the infection status of the patients in residence. In rooms with no positive *Clostridium difficile* patient, there was a low rate (8%) of *Clostridium difficile* positive assays of environmental sites. Once a positive *Clostridium difficile* patient was admitted into a room, the environmental rate increased and was 29% if the patient was an asymptomatic carrier and 49% if the patient developed diarrhea [37]. Once the environmental isolates were immunoblot typed and compared to the suspected source patient's isolate, 83% were of the identical immunoblot type. Increases in environmental contamination by *Clostridium difficile* have been documented by several studies once an infected patient is admitted [17, 25] and studies have also documented that the environmental isolate was of the same strain as the source patient's isolate [22, 25, 58].

Clostridium difficile may also be carried and spread by the hands of hospital personnel. Studies have shown that 1.5-59% of hospital personnel have *Clostridium difficile* present on hand surfaces [13, 17, 25, 37], although not all studies find hand contamination [54]. Hand imprint cultures of hospital personnel caring for a *Clostridium difficile* positive patient have documented 59% of the hands became positive after contact and 95% of the hand isolates were of the same immunoblot type as the suspected source patient [37]. Rarely have hospital personnel been shown to be carriers of *Clostridium difficile* from stool cultures [17, 25].

Roommates may also be a source (and target) of *Clostridium difficile*. In a study of 399 hospitalized patients on one general medicine ward, 92 were exposed to an infected roommate and 25% of the roommates acquired *Clostridium difficile*. Of 23 roommates who subsequently acquired *Clostridium difficile*, 20 (87%) were of the same immunoblot type as the suspected source patient [37]. Another study found the frequency of *Clostridium difficile* rose from 14% to 33% after exposure to a *Clostridium difficile* infected roommate [54].

The source or vehicle for outbreaks of CDAD is rarely documented due to the lag time between the initial spread and the recognition of an outbreak of CDAD. In several studies, the vehicle has been documented (Table 2) and has included both environmental sites and hospital personnel.

Table 2. Implicated sources and vehicles in nosocomial outbreaks of *Clostridium difficile*-associated disease

Year	Type of patients	Number	Source or vehicle	Reference
1979	Inpatients	66	Sigmoidoscope	[39]
1982	Neonates	Nd	Hosp personnel	[27]
1983	Inpatients	2	Commode chair	[55]
1987	Inpatients	23	Hosp personnel	[46]
1988	Inpatients	10	Sluice room	[58]
1990	Gen. Med.	21	Commode chair	[37]

Nd = Not determined

Sub-populations at risk

Clostridium difficile outbreaks have been reported in general medicine patients, surgical patients and patients at nursing homes since the 1980's [2, 3, 10, 19, 37, 45, 54, 60, 61, 63]. More recently, *Clostridium difficile* has spread into patient

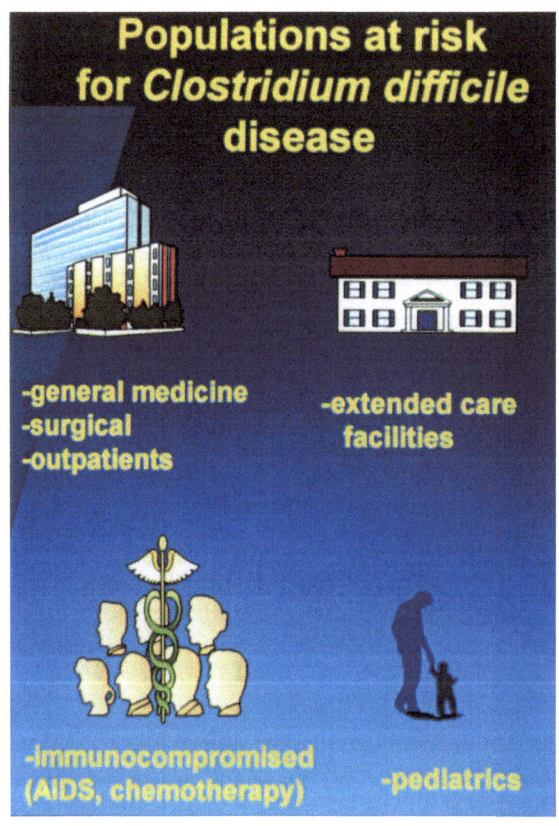

Fig. 3. Populations at risk for *C. difficile* disease

populations which have been assumed to be of low risk for CDAD (Fig. 3). *Clostridium difficile* outbreaks have been found in pediatric populations [4, 7, 9] and in neonates [23]. Patients receiving cancer chemotherapy have reported increased rates of *Clostridium difficile* [7, 15, 53]. Recently, patients with AIDS (acquired immunodeficiency disease syndrome) have been found to have CDAD or have experienced an outbreak of CDAD [1, 20, 29].

Risk factors for *Clostridium difficile*

Numerous risk factors have been found for *Clostridium difficile* disease in studies using univariate analyses. These risk factors include advanced age, female gender, severe underlying diseases, number and types of antibiotics, number and types of medications, gastrointestinal procedures (enemas, endoscopy, surgery), nasogastric alimentation, renal disease and length of stay in the hospital [24, 40, 49, 52, 63, 64]. Because most of these variables are closely related, it is difficult to obtain an independent estimate of risk for a specific factor. The use of multivariate models allows an estimation of the relative risk of a specific factor while simultaneously adjusting the influences of other factors. Thus, multivariate models can provide an estimate of the risk for specific factors such as the number of antibiotics while adjusting for differing lengths-of-stay. Studies which utilize multivariate models have found that antibiotics (penicillins, cephalosporins, clindamycin) have significant risk for CDAD, in addition to age and procedures or medications which disrupt normal colonic flora. Brown et al studied 74 patients and found four risk factors for CDAD: age over 65 years (RR = 14.1, 95% C.I. 1.4, 141), stay in the intensive care unit (RR = 39.2, 95% C.I. 2.2, 713), antibiotic use over 10 days (RR = 16.1, 95% C.I. 2.2, 117) and gastrointestinal procedures (RR = 23.2, 95% C.I. 2.1, 255) which included nasogastric tubes, enemas, or upper intestinal series [6]. In a prospective study of 399 general medicine patients, multivariate analysis revealed five risk factors for CDAD after adjusting for age and severity of underlying disease: cephalosporin use less than one week (RR = 2.07, 95% C.I. 1.12, 3.84), penicillin use for one to two weeks (RR = 3.41, 95% C.I. 1.48, 7.86), enemas (RR = 3.26, 95% C.I. 1.51, 7.02), intestinal stimulants (RR = 3.06, 95% C.I. 1.67, 5.60), and stool softeners (RR = 1.74, 95% C.I. 1.02, 3.00). Factors which were not found to be significant for CDAD included gender, antacid use, nasogastric tube exposure, surgery, renal conditions, other antibiotics or medications or immunocompromised states [36]. Nelson et al studied 65 patients at a Veteran's Medical Center and found 3 risk factors using multivariate analysis: clindamycin (RR = 7.2, 95% C.I. 1.5, 35.6), second and third generation cephalosporins (RR = 8.3, 95% C.I. 1.4, 48.9) and use of multiple antibiotics (RR = 18.7, 95% C.I. 4.1, 85.8). This study did not find that age, gender, length of stay, APACHE scores, medications or upper gastrointestinal procedures were significantly associated with CDAD [45]. In 57 HIV positive patients, only two risk factors for CDAD were found: clindamycin (OR = 42.0, 95% C.I. 2.0, 813) and length of stay (OR = 3.6, 95% C.I. 1.0, 13.0) but no other factors were significant [20]. The common thread in these risk factors is the impact the variables have on the normal colonic flora. Factors which disrupt the «colonization resistance» or barrier effect of normal intestinal flora can be expected to increase the susceptibility of patients to CDAD.

Control measures for nosocomial outbreaks

For the past two decades, physicians and infection control practitioners have been struggling to determine effective methods of controlling nosocomial outbreaks of CDAD. Once the routes of transmission were determined, strategies to interrupt these routes were studied. Because *Clostridium difficile* and its spores are resistant to the usual ammoniacal solutions, disinfectants with either alkaline glutaraldehyde (0.03-2%) or alkaline hypochlorite (500-1600 ppm) solutions have been studied [22, 56]. Both of these types of disinfectants were successful in reducing rates of CDAD (Table 3). To prevent cross infections due to hand carriage, studies using bactericidal handwashing soaps or vinyl gloves were effective in reducing *C. difficile* [21, 37, 58]. If equipment is difficult to disinfect, studies found replacement by disposable, single-use equipment lowered CDAD rates [5, 58].

Table 3. Interventions for the control of nosocomial outbreaks of *Clostridium difficile-associated disease* (CDAD)

Patient population	Type of intervention	Frequency of positive cultures		Reference
		Before	After	
Environmental sites	Hypochlorite (500 ppm)	31%	16%	[22]
Adult inpatients	Education, vinyl glove	7.7/1000	1.5/1000	[21]
Environmental site	Environ. decontamination, handwashing, disposable equipment	21/150	0	[58]
Hospital staff	Bactericidal handwashing soap	88%	14%	[37]
Elderly patients	Increase infection control practice	13%	6.3%	[30]
Surgery patients	Enteric precautions, surveillance, terminal room, disinfection, cohorting, early treatment	1.5/1000	0.3/1000	[56]
4 wards	Infect control education	15/mon	5/mo	[44]
Inpatients	Single use disposable thermometers	2.7/1000	1.8/1000	[5]
VAMC*	Clindamycin	7.7/mon	1.9/mon	[48]

*VAMC = Veteran's Administration Medical Center

Another strategy for controlling and preventing nosocomial outbreaks of CDAD was to focus on the susceptibility state of the patient. Antibiotic utilization programs were instituted to evaluate the patterns of use of high risk antibiotics and to limit their use. Clindamycin restriction was effective in reducing the rate of CDAD from 7.7 per month to 1.9 per month [48]. At a Veteran's Administration Hospital, the antibiotic restriction program was discontinued which led to a 20% increase in beta-lactam antibiotic use and a subsequent doubling of their CDAD incidence [41].

The most effective programs for the control of outbreaks have been strategies which combined several avenues for the interruption of transmission. Brown et al found that control measures including early isolation and treatment of CDAD cases and clindamycin restriction reduced the rate of CDAD from 2.25% to 0.74% [6]. Cartmill et al found that control measures including intensive staff education about CDAD, increased vigilance, strict enteric precautions, rigorous cleaning procedures, restriction of staff and patient transfers and antibiotic restriction lead to a decrease in CDAD cases [10]. Despite these efforts, some hospitals have reported a failure to significantly reduce the sustained persistence of CDAD at their institutions [11, 32, 45, 47, 51].

Recommended guidelines

Prevention of CDAD outbreaks relies on several different guidelines. In order to recognize the existence of an outbreak, surveillance problems should be instituted in order to detect a higher than normal frequency of antibiotic-associated diarrhea cases. Prompt diagnosis and treatment of patients with symptomatic CDAD is recommended to limit the transmission of the organism to other patients, personnel and the environment. Effective environmental decontamination procedures should be standardized and disposal single-use equipment should be used especially for equipment exposed to fecal contents. Infection control programs are effective (education of enteric precautions, use of gloves, handwashing techniques, handling of soiled linens and equipment). Hospitals may benefit from antibiotic utilization programs which analyze and restrict the use of high risk antibiotics.

References

1. Barbut F, Mario N, Meyohas MC, et al (1994) Investigation of a nosocomial outbreak of *Clostridium difficile*-associated diarrhoea among AIDS patients by random amplified polymorphic DNA (RAPD) assay. J Hosp Infect 26: 181-189
2. Bender BS, Bennett R, Laughon BE, et al (1986) Is *Clostridium difficile* endemic in chronic-care facilities? Lancet i: 11-13
3. Bennett GC, Allen E, Millard PH (1984) *Clostridium difficile* diarrhoea: A highly infectious organism. Age Ageing 13: 363-366
4. Bowen KE, McFarland LV, Greenberg RN, et al (1995) Isolation of *Clostridium difficile* (CD) in a university hospital: a 2-year study (April 1990-April 1992). Clin Infect Dis 20 (suppl 2): S261-262

5. Brooks SE, Veal RO, Kramer M, et al (1992) Reduction in the incidence of *Clostridium difficile*-associated diarrhea in an acute care hospital and a skilled nursing facility following replacement of electronic thermometers with single-use disposables. Infect Control Hosp Epidemiol 13: 98-103

6. Brown E, Talbot GH, Axelrod P, et al (1990) Risk factors for *Clostridium difficile* toxin-associated diarrhea. Infect Control Hosp Epidemiol 11: 283-290

7. Brunetto AL, Pearson AD, Craft AW, et al (1988) *Clostridium difficile* in an oncology unit. Arch Dis Child 63:979-981

8. Burdon DW, Mogg A, Alexander-Williams J, et al (1980) Epidemiology of antibiotic-associated colitis. Current Chemotherapy and Infectious Diseases 2: 953-954

9. Buts J-P, Corthier G, Delmée M (1993) *Saccharomyces boulardii* for *Clostridium difficile*-associated enteropathies in infants. J Pediatr Gastroenterol Nutr 16: 419-425

10. Cartmill TD, Panigrahi H, Worsley MA, et al (1994) Management and control of a large outbreak of diarrhoea due to *Clostridium difficile*. J Hosp Infect 27: 1-15

11. Cartmill TDI, Shrimpton SB, Panigrahi H, et al (1992) Nosocomial diarrhoea due to a single strain of *Clostridium difficile*: A prolonged outbreak in elderly patients. Age Ageing 21: 245-249

12. Clabots CR, Johnson S, Olson MM, et al (1992) Acquisition of *Clostridium difficile* by hospitalized patients: Evidence for colonized new admissions as a source of infection. J Infect Dis 166: 561-567

13. Cudmore MA, Silva J Jr, Fekety R, et al (1982) *Clostridium difficile* colitis associated with cancer chemotherapy. Arch Intern Med 142: 333-335

14. Delmée M, Vandercam B, Avesani V, et al (1987) Epidemiology and prevention of *Clostridium difficile* infections in a leukemia unit. Eur J Clin Microbiol 6: 623-627

15. Delmée M, Bulliard G, Simon G (1986) Application of a technique for serogrouping *Clostridium difficile* in an outbreak of antibiotic-associated diarrhoea. J Infect 13: 5-9

16. Eriksson S, Aronsson B (1989) Medical implications of nosocomial infection with *Clostridium difficile*. Scand J Infect Dis 21: 733-734

17. Fekety R, Kim K-H, Brown D, et al (1981) Epidemiology of antibiotic-associated colitis. Isolation of *Clostridium difficile* from the hospital environment. Am J Med 70: 906-908

18. Foulke GE, Silva J Jr (1989) *Clostridium difficile* in the intensive care unit: Management problems and prevention issues. Crit Care Med 17: 822-826

19. Gerding DN, Olson MM, Peterson LR, et al (1986) *Clostridium difficile*-associated diarrhea and colitis in adults. A prospective case-controlled epidemiologic study. Arch Intern Med 146: 95-100

20. Hutin Y, Molina JM, Casin I, et al (1993) Risk factors for *Clostridium difficile*-associated diarrhoea in HIV-infected patients. AIDS 7: 1441-1447

21. Johnson S, Gerding DN, Olson MM, Weiler MD, Hughes RA, Clabots CR, Peterson LR (1990) Prospective, controlled study of vinyl glove use to interrupt *Clostridium difficile* nosocomial transmission. Am J Med 88: 137-140

22. Kaatz GW, Gitlin SD, Schaberg DR, et al (1988) Acquisition of *Clostridium difficile* from the hospital environment. Am J Epidemiol 127: 1289-1294

23. Kato H, Kato N, Watanabe K, et al (1994) Application of typing by pulsed-field gel electrophoresis to the study of *Clostridium difficile* in a neonatal intensive care unit. J Clin Microbiol 32: 2067-2070

24. Keighley MRB, Burdon DW, Alexander-Williams J, et al (1978) Diarrhoea and pseudomembranous colitis after gastrointestinal operations. Lancet ii: 8101-8102

25. Kim K-H, Fekety R, Batts DH, et al (1981) Isolation of *Clostridium difficile* from the environment and contacts of patients with antibiotic-associated colitis. J Infect Dis 143: 42-49

26. Kofsky P, Rosen L, Reed J, et al (1991) *Clostridium difficile* - a common and costly colitis. Dis Colon Rectum 34: 244-248

27. Larson HE, Barclay Fe, Honour P, et al (1982) Epidemiology of *Clostridium difficile* in infants. J Infect Dis 146: 727-733

28. Lima NL, Guerrant RL, Kaiser DL, et al (1990) A retrospective cohort study of nosocomial diarrhea as a risk factor for nosocomial infection. J Infect Dis 161: 948-952

29. Lu SS, Schwartz JM, Simon DM, et al (1994) *Clostridium difficile*-associated diarrhea in patients with HIV positivity and AIDS: a prospective controlled study. Am J Gastroenterol 89: 1226-1229

30. Magee JT, Brazier JS, Hosein IK, et al (1993) An investigation of a nosocomial outbreak of *Clostridium difficile* by pyrolysis mass spectrometry. J Med Microbiol 39: 345-351

31. Malamou-Ladas H, O'Farrell S, Nash JQ, et al (1983) Isolation of *Clostridium difficile* from patients and the environment of hospital wards. J Clin Pathol 36: 88-92

32. Manian FA, Meyer L (1995) CDAD Rates (Letter). Infect Control Hosp Epidemiol 16: 63-64

33. Marts BC, Longo WE, Vernava AM III, et al (1994) Patterns and prognosis of *Clostridium difficile* colitis. Dis Colon Rectum 37: 837-845

34. McFarland LV (1995) *Clostridium difficile*-associated disease. In: Surawicz C, Owen RL (eds) Gastrointestinal and hepatic infections. WB Saunders Co, Philadelphia, pp 153-175

35. McFarland LV (1993) Diarrhea acquired in the hospital. In: Giannella R (ed) Gastroenterol Clin North Am 22: 563-577

36. McFarland LV (1990) *Clostridium difficile*: epidemiology and potential reservoirs. In: Rambaud JC, Ducluzeau R (eds) *Clostridium difficile*-associated intestinal diseases. Springer, Paris, pp 69-80

37. McFarland LV, Mulligan ME, Kwok RYY, Stamm WE (1989) Nosocomial acquisition of *Clostridium difficile* infection. N Engl J Med 320: 204-210

38. McKay I, Coia JE, Poxton IR (1989) Typing of *Clostridium difficile* causing diarrhoea in an orthopaedic ward. J Clin Pathol 42: 511-515

39. Mogg GAG, Keighley MRB, Burdon DW, et al (1979) Antibiotic-associated colitis - a review of 66 cases. Br J Surg 66: 738-742

40. Morris JB, Zollinger RM, Stellato TA (1990) Role of surgery in antibiotic-induced pseudomembranous enterocolitis. Am J Surg 160: 535-539

41. Mulligan M, Berman S, Antony T, et al (July 1994) An outbreak of severe disease due to *Clostridium difficile* following decreased antibiotic restriction. (Abstract) Anaerobic Society Americas Meeting, Marina Del Ray, CA

42. Mulligan ME, George WL, Rolfe RD, et al (1980) Epidemiological aspects of *Clostridium difficile*-induced diarrhea and colitis. Am J Clin Nutr 33 (11 suppl): 2533-2538

43. Mulligan ME, Rolfe RD, Finegold SM, et al (1979) Contamination of a hospital environment by *Clostridium difficile*. Curr Microbiol 3: 173-175

44. Nath SK, Thornley JH, Kelly M, et al (1994) A sustained outbreak of *Clostridium difficile* in a general hospital: persistence of a toxigenic clone in four units [see comments]. Infect Control Hosp Epidemiol 15: 382-389

45. Nelson DE, Auerbach SB, Baltch AL, et al (1994) Epidemic *Clostridium difficile*-associated diarrhea: Role of second- and third-generation cephalosporins. Infect Control Hosp Epidemiol 15: 88-94

46. Nolan NPM, Kelly CP, Humphreys JFH, et al (1987) An epidemic of pseudo-membranous colitis: Importance of person to person spread. Gut 28: 1467-1473

47. Olson MM, Shanholtzer CJ, Lee JT Jr, et al (1994) Ten years of prospective *Clostridium difficile*-associated disease surveillance and treatment at the Minneapolis VA Medical Center, 1982-1991. Infect Control Hosp Epidemiol 15: 371-381

48. Pear SM, Williamson TH, Bettin KM, et al (1994) Decrease in nosocomial *Clostridium difficile*-associated diarrhea by restricting clindamycin use. Ann Intern Med 120: 272-277

49. Pierce PF Jr, Wilson R, Silva J Jr, et al (1982) Antibiotic-associated pseudo-membranous colitis: An epidemiologic investigation of a cluster of cases. J Infect Dis 145: 269-274

50. Poxton IR, Aronsson B, Mollby R, et al (1984) Immunochemical fingerprinting of *Clostridium difficile* strains isolated from an outbreak of antibiotic-associated colitis and diarrhoea. J Med Microbiol 17: 317-324

51. Riley TV, O'Neill GL, Bowman RA, et al (1994) *Clostridium difficile*-associated diarrhoea: epidemiological data from Western Australia. Epidemiol Infect 113: 13-20

52. Rosenberg JM, Walker M, Welch JP, et al (1984) *Clostridium difficile* colitis in surgical patients. Am J Surg 147: 486-491

53. Sakai C, Kumagai K, Takagi T, et al (1993) An epidemic of *Clostridium difficile* colitis in patients with cancer: role of cancer chemotherapy and nosocomial infection in the pathogenesis. Gan To Kagaku Ryoho 20: 2413-2416

54. Samore MH, Bettin KM, DeGirolami PC, et al (1994) Wide diversity of *Clostridium difficile* types at a tertiary referral hospital. J Infect Dis 170: 615-621

55. Savage AM, Alford RH (1983) Nosocomial spread of *Clostridium difficile*. Infection Control 4: 31-33

56. Struelens MJ, Maas A, Nonhoff C, et al (1991) Control of nosocomial transmission of *Clostridium difficile* based on sporadic case surveillance. Am J Med 91 (suppl 3B): S138-144

57. Tabaqchali S, Holland D, O'Farrell S, et al (1984) Typing scheme for *Clostridium difficile*: Its application in clinical and epidemiological studies. Lancet i: 935-958

58. Testore GP, Pantosti A, Cerquetti M, et al (1988) Evidence for cross-infection in an outbreak of *Clostridium difficile*-associated diarrhoea in a surgical unit. J Med Microbiol 26: 125-128

59. Thomas DR, Bennett RG, Laughon BE, et al (1990) Postantibiotic colonization with *Clostridium difficile* in nursing home patients. J Am Geriatr Soc 38: 415-420

60. Vautrin AC, Enck S, Antoniotti G, et al (1993) Survey with three epidemiological markers after 22 cases of diarrhea caused by *Clostridium difficile* in a geriatric hospital. Pathol Biol (Paris) 41: 421-427

61. Walker KJ, Gilliland SS, Vance-Bryan K, et al (1993) *Clostridium difficile* colonization in residents of long-term care facilities: prevalence and risk factors. J Am Geriatr Soc 41: 940-946

62. Wust J, Sullivan NM, Hardegger U, et al (1982) Investigation of an outbreak of anti-biotic-associated colitis by various typing methods. J Clin Microbiol 16: 1096-1101

63. Yee J, Dixon CM, McLean APH, et al (1991) *Clostridium difficile* disease in a department of surgery. Arch Surg 126: 241-246

64. Zimmerman RK (1991) Risk factors for *Clostridium difficile* cytotoxin-positive diarrhea after control for horizontal transmission. Infect Control Hosp Epidemiol 12: 96-100

Prevalence and pathogenicity of *Clostridium difficile*. A French multicenter study

Jean-Claude Petit,
The study group* on *Clostridium difficile* prevalence

SUMMARY

Clostridium difficile is recognized as being a causal agent in nosocomial diarrhea in adults, occurring during or after antibiotic treatment. However its prevalence in the stools of hospital patients received in bacteriological laboratories is not precisely established. In order to establish this prevalence, the stools of all hospital patients received by 11 laboratories throughout France were routinely assayed for *C. difficile*, from January to July 1993. The prevalence of *C. difficile* in those patients for whom a stool culture had been requested by a doctor (termed « cases ») was compared with that from a control group in whom stool culture had been required by the laboratory although not prescribed by the doctor in charge of the patient. This control group was stratified according to referring team, age, and length of stay. In all, 3921 cases and 229 controls were included. Recorded data in each patient included age, sex, referring team, length of hospital stay, and gross appearance of the stool. In addition, if *C. difficile* was found, more detailed clinical information was sought from the referring clinician. Each strain of *C. difficile* was serotyped and studied for toxigenicity.

The prevalence of *Clostridium difficile* was twice as high in the cases as in the control group (9.7% vs. 4.8%, $p < 0.01$). As regards to gross appearance, the prevalence of *C. difficile* was four times higher in the diarrheal stools (in both cases and controls) than that seen in the non-diarrheal stools of the controls [11.5% (cases) and 11.6% (controls) vs 3.3%]. The prevalence of *C. difficile* was higher in the stools of patients aged over 65, those hospitalized for more than one week, those from intensive care units, long stay patients and those from infectious diseases units. Furthermore, 52% of *C. difficile* carriers had co-existing chronic or severe illness. The strains of *C. difficile* were more often toxin producers when isolated from diarrheal stools (72% vs. 55%) or from patients with clinical diarrhea (71% vs. 51%). Analysis of the various serotypes showed that type D (never toxigenic in our study) was statistically isolated less often from diarrheal stools and from those from patients with clinical diarrhea.

The results of this multicenter prevalence study from 11 french laboratories over a six month period strengthen the role of *C. difficile* in the etiology of infectious diarrhea in hospital patients. *C. difficile* should be routinely sought in diarrheal stools from these patients.

Prévalence et pathogénicité de *Clostridium difficile* : une étude multicentrique française

RÉSUMÉ

C. difficile est reconnu comme un agent responsable de diarrhées nosocomiales chez l'adulte, survenant pendant ou au décours d'un traitement antibiotique. Cependant sa prévalence dans les selles de patients hospitalisés, reçues par les laboratoires de bactériologie, n'est pas précisément établie. Dans le but de déterminer cette prévalence, toutes les selles de patients hospitalisés reçues par 11 laboratoires répartis en France ont fait l'objet d'une recherche systématique de *C. difficile*, de janvier à juillet 93. La prévalence de *C. difficile* chez les patients pour lesquels une coproculture avait été prescrite par le médecin (appelés « cas ») a été comparée avec la prévalence de *C. difficile* dans un groupe de témoins aléatoires pour lesquels les coprocultures ont été demandées par le laboratoire, bien que non prescrites par le médecin responsable du patient. Ce groupe témoin a été stratifié par rapport au groupe de référence en fonction de l'unité de soins, de l'âge du patient et de la durée du séjour. Au total, 3921 cas et 229 témoins ont été inclus. Les données enregistrées pour chaque patient ont été l'âge, le sexe, l'unité de soins, la durée d'hospitalisation et l'aspect macroscopique de la selle. De plus, quand *C. difficile* a été isolé, des informations cliniques plus détaillées ont été recueillies auprès du clinicien concerné. Chaque souche de *C. difficile* a fait l'objet d'un sérotypage et d'une étude de toxinogenèse.

La prévalence de *C. difficile* est deux fois plus élevée dans les coprocultures prescrites (« cas ») que dans le groupe témoin (9,7 % versus 4,8 % ; p < 0,01). En ce qui concerne l'aspect macroscopique, la prévalence de *C. difficile* est quatre fois plus élevée dans les selles diarrhéiques (chez les cas et les témoins) que celle observée dans les selles non diarrhéiques des témoins (11,5 % chez les cas et 11,6 % chez les témoins, versus 3,3 %). La prévalence de *C. difficile* est plus élevée dans les selles des patients ayant plus de 65 ans, chez ceux hospitalisés depuis plus d'une semaine, chez les patients qui sont en unités de soins intensifs, de long séjour ou de maladies infectieuses. De plus, 52 % des porteurs de *C. difficile* ont une pathologie sévère ou chronique sous-jacente. Les souches de *C. difficile* sont plus souvent productrices de toxines quand elles sont isolées de selles diarrhéiques (72 % versus 55 %) ou de selles de patients présentant une diarrhée clinique (71 % versus 51 %). L'analyse des différents sérotypes montre que le sérogroupe D, qui n'est jamais toxinogène dans notre étude, est statistiquement moins souvent isolé de selles diarrhéiques ou des selles de patients présentant une diarrhée clinique.

Les résultats de cette étude multicentrique de prévalence réalisée dans 11 laboratoires français sur une période de 6 mois renforcent l'hypothèse du rôle de *C. difficile* dans l'étiologie des diarrhées infectieuses chez les patients hospitalisés. *C. difficile* devrait donc faire l'objet d'une recherche systématique dans les selles diarrhéiques de ces patients.

Prävalenz und Pathogenität von *Clostridium difficile*
Französische Multicenterstudie

ZUSAMMENFASSUNG

C. difficile ist als kausaler Auslöser nosokomialer Diarrhöen bei Erwachsenen während oder nach einer Antibiotikabehandlung bekannt. Bisher jedoch wurde die Prävalenz dieses Keims im Stuhlproben stationärer Patienten nicht in bakteriologischen Labors exakt ermittelt.

Zur Bestimmung dieser Prävalenz wurden zwischen Januar und Juli 1993 alle eingesandten Stuhlproben von stationären Patienten in 11 Labors in ganz Frankreich routinemäßig auf *C. difficile* untersucht. Die Prävalenz von *C. difficile* bei Patienten, für die eine Stuhluntersuchung angefordert worden war ("Fälle" genannt), wurde verglichen mit einer Kontrollgruppe, für die Stuhluntersuchungen zwar nicht vom behandelnden Arzt angefordert, jedoch vom Labor durchgeführt worden waren. Diese Kontrollgruppe wurde nach überweisender Stelle, Alter und Dauer des stationären Aufenthaltes stratifiziert. Insgesamt wurden 3.921 Fälle und 229

Kontrollen erfaßt. Bei jedem Patienten wurden Angaben zu Alter, Geschlecht, Herkunft der Anforderung, Zeitraum bis zur stationären Aufnahme und das makroskopische Erscheinungsbild des Stuhls festgehalten. War *C. difficile* nachweisbar, wurde der behandelnde Arzt darüber hinaus um detaillierte klinische Informationen gebeten. Jeder *C. difficile*-Stamm wurde serologisch typisiert und auf seine Toxinogenität untersucht.

Die Prävalenz von *C. difficile* war bei den "Fällen" doppelt so hoch wie bei den Kontrollen (9,7 % gegenüber 4,8 %, p < 0,01). Im Hinblick auf das makroskopische Aussehen der Stühle war die Prävalenz von *C. difficile* bei den "Fällen" ebenso wie bei Kontrollen in durchfälligen Stühlen viermal so hoch wie in nicht durchfälligen Stühlen (11,5 % {Fälle} bzw. 11,6 % {Kontrollen} gegenüber 3,3 %). Die Prävalenz von *C. difficile* war erhöht im Stuhl von Patienten über 65 Jahre, länger als eine Woche stationär behandelten Patienten, bei Patienten auf Intensivstationen, Langzeitpatienten und Patienten mit Infektionskrankheiten. Darüber hinaus litten 52 % der *C. difficile*-Carrier an einer begleitenden chronischen oder schweren Krankheit. Die *C. difficile*-Stämme waren häufiger toxinogen, wenn sie aus durchfälligen Stühlen (72 % vs. 55 %) oder aus Stuhlproben von Patienten mit klinisch manifester Diarrhoe (71 % vs. 51 %) isoliert wurden. Die Analyse der verschiedenen Serotypen ergab, daß Typ D (der in unserer Studie niemals toxinogen war), statistisch weniger häufig aus durchfälligen Stühlen oder aus Stuhlproben von Patienten mit klinisch manifester Diarrhoe isoliert wurde.

Die Ergebnisse dieser über einen Zeitraum von sechs Monaten multizentrisch in 11 französischen Labors durchgeführten Prävalenzuntersuchung unterstreicht die Bedeutung von *C. difficile* in der Ätiologie infektiöser Diarrhöen bei stationär behandeiten Patienten. Durchfällige Stuhlproben dieser Patienten soliten routinemäßig auf *C. difficile* untersucht werden.

*Members of the *C. difficile* prevalence study group (in alphabetical order):

B. De Barbeyrac, Service de Bactériologie, Groupe hospitalier Pellegrin-Tripode, Bordeaux, France; F. Barbut, Service de Bactériologie-Virologie, Hôpital Saint-Antoine, Paris, France; C. Bebear, Service de Bactériologie, Groupe hospitalier Pellegrin-Tripode, Bordeaux, France; P. Bernasconi, Laboratoires Biocodex, Paris, France; Y. Boussougant, Service de Bactériologie, Hôpital Louis Mourier, Colombes, France; M. Cerf, Service d'Hépato-Gastro-Entérologie, Hôpital Louis Mourier, Colombes, France; Y. Charpak, Société Eval, Paris, France; A. Collignon, Service de Bactériologie, Hôpital Jean Verdier, Bondy, France; G. Corthier, Laboratoires d'Ecologie et Physiologie du Système Digestif, Institut National de Recherche Agronomique, Jouy-en-Josas, France; S. Duborgel, Service de Bactériologie, Faculté de Médecine, Grenoble, France; T. Fosse, Laboratoire de Bactériologie, Hôpital Pasteur, Nice, France; M.E. Le Guern, Laboratoires Biocodex, Paris, France; P. Le Noc, Service de Bactériologie, Faculté de Médecine, Grenoble, France; H. Monteil, Institut de Bactériologie, Université Louis Pasteur, Strasbourg, France; V. Schue, Institut de Bactériologie, Université Louis Pasteur, Strasbourg, France; A. Sédallian, Service de Bactériologie, Centre Hospitalier, Annecy, France; S. Tigaud, Service de Microbiologie, Hôpital de la Croix-Rousse, Lyon, France; A. Trévoux, Service de Bactériologie, Hôpital Emile Muller, Mulhouse, France; F. Tytgat, Service de Bactériologie, Centre Hospitalier Sainte-Catherine, Saverne, France.

Introduction

Clostridium difficile is responsible for pseudomembranous colitis and diarrhea occurring during or after treatment with antibiotics [11, 15, 20]. The asymptomatic carrier rate among hospital patients on admission is from 1.4 to 11%, and the rate of nosocomial transmission varies from 8.5 to 32.2% [4, 6, 12, 14, 16, 19].

In spite of this, routine search for *C. difficile* and/or its toxins in the stools of hospital patients is not currently practised in France. It seems that there are various reasons for this, including ignorance on the part of doctors regarding this organism and lack of facilities for its identification in laboratories. More probably, it is the significance of the isolation of a toxigenic strain of *C. difficile* other than in cases of pseudomembranous colitis, which is still debatable. In fact, *C. difficile* is still not generally recognised as a «classical» enteropathogen which would justify routine search for it in the stools, in view of the high asymptomatic carrier rate [4, 16, 19], and the lack of well established correlation between pathogenicity, toxin production and serotype [18].

In view of these many unresolved questions, a large multicenter prospective study was undertaken, of the prevalence of *C. difficile* in the stools of hospital patients.

The aims were: (1) to establish the prevalence of *C. difficile* in stools received by hospital microbiology laboratories; (2) to establish the relation between the finding of *C. difficile* and the gastrointestinal condition of the patient; (3) to characterize the strains of *C. difficile* isolated, by measurement of production of toxins A and B, and by serotyping.

Materials and methods

Conduct of the study

A routine search for *C. difficile* was undertaken in all stools received from hospital patients, in 11 hospital microbiology laboratories throughout France, over the six month period from January to July 1993. Specimens from departments of oncology and hematology were excluded, because of the routine study of stools from these patients for non-digestive reasons, and also those from children under two years of age, as they are frequent asymptomatic carriers [3, 9, 13]. The names were registered, in order to avoid examining the same patient twice.

Cases and controls

In order to assess the relation between the presence of *C. difficile* and gastro-intestinal tract pathology, the prevalence of *C. difficile* in those patients in whom a stool culture had been requested by the doctor (the « cases ») was compared with that where the laboratory had carried out the stool culture although it had not been asked for by the clinician in charge. This control group was stratified according to referring team, age, length of hospital stay between admission and date of collection of the specimen. Bearing in mind that it was hoped to include 4000 cases, the number of controls was estimated to be 250 in order to ensure a type 1 error rate of 5% and a power of 90%.

Collection of data

For both case and control groups, the age, sex, referring team, length of hospital stay, and gross appearance of the stool (diarrheal or not) were recorded. If *C. difficile* was isolated, more detailed clinical information was sought, from colleagues or from the case notes, to include the reason for admission, the presence of clinical diarrhea (defined by three or more soft or liquid stools per day), the date of onset and duration of the diarrhea, antibiotic treatment received in the preceding month, and other known risk factors such as gastrointestinal tract decontamination, cancer chemotherapy, immunosuppressive treatment, associated chronic or severe illness, enteral feeding, and HIV status.

Bacteriological examination

A standard culture technique for *C. difficile* was adopted by all 11 laboratories, who were already familiar with it. This employed a selective agar culture medium containing cefoxitin (10 mg/l) and cycloserine (250 mg/l) (*C. difficile* agar bioMérieux, La Balme-les-Grottes, France) incubated for 48 hours under anaerobic conditions. *C. difficile* was identified using biochemical reactions (Rapid ID 32A, bioMérieux, La Balme-les Grottes, France).

For other enteropathogens, a routine search was carried out for Salmonella, Shigella, Campylobacter and Yersinia, together with other organisms chosen freely by the laboratory.

The strains of *C. difficile* were sent to a reference laboratory (G. Corthier, National Institute for Agronomic Research, Jouy-en-Josas) for toxin studies and serotyping. Identification of toxins A and B was carried out using the supernatant from a Trypticase Yeast Glucose culture broth (Diagnostics Pasteur, Marne-la-Coquette, France) incubated anaerobically for 5 days. Toxin A production was studied by an ELISA technique using a monoclonal anti-toxin A antibody, and toxin B production by observing the pathological effects on CHO-K1 cells.

The serotypes were established by Delmée's method, recently modified into an ELISA technique [5, 8] using polyclonal antibodies (A1, A5, A8, A9, A10, C, D, F, G, H, K).

Statistical analysis

This was carried out using the SAS™ software. The tests applied were Chi squared (with Yates' correction when sample sizes were small) for qualitative variables, and analysis of variance for quantitative variables. Odds ratios were calculated to compare the cases and controls for the presence of *C. difficile* (Epi-Info™ software); odds ratios were also calculated for the subgroups according to the appearance of the stools (diarrheal or not). Regression analysis was carried out in order to determine the correlation between the presence of *C. difficile* and other variables.

Results

Comparison between cases and controls

The two groups were matched as regards age, sex, referring department and time since admission.

Prevalence of C. difficile in cases and controls

C. difficile was found twice as often in the cases as the controls (Table 1). As regards to gross consistency of the stool, it was observed that the prevalence of *C. difficile* was four times higher in loose stools in both cases and controls, than in controls without diarrheal stools (Table 1).

Table 1. Prevalence of *C. difficile* in cases and controls

Status	Prevalence of C. difficile (%)	Odds ratio	p value
Overall analysis			
Controls (n = 229)	4.8	1	-
Cases (n = 3921)	9.7	2.13	< 0.01*
Analysis by sub-group***			
Controls			
without diarrheal stools (n = 184)	3.3	1	-
with diarrheal stools (n = 43)	11.6	3.9	< 0.05**
Cases			
without diarrheal stools (n = 1834)	7.7	2.5	< 0.05*
with diarrheal stools (n = 2066)	11.5	3.8	< 0.001*

* Pearson's chi^2 test
** Fischer's exact test
*** missing data for the variable «consistency of the stool»

Prevalence of C. difficile in the cases according to different variables

The prevalence of *C. difficile* was not statistically related to the referring team, though it was observed that the prevalence was much higher in patients from intensive care units, in long stay patients and in those from infectious diseases units (Table 2).

There was no correlation between sex and *C. difficile* prevalence. However the prevalence increased with age, to a slight degree from childhood to 65, but thereafter more rapidly.

The time from date of admission to stool culture showed that the prevalence of *C. difficile* in the stools was higher in patients when the request for culture had been made either in the first 24 hours after admission, or after the first week.

Table 2. Prevalence of *C. difficile* according to different variables (cases only)

Variables	Prevalence of *C. difficile* (%)	p value
Referring department		
Infectious diseases (n = 707)	11	NS
Internal medicine (n = 1732)	10	
Long stay & Geriatrics (n = 176)	13	
Pediatrics (n = 360)	6	
Adult surgery (n = 466)	9	
ICU (n = 433)	11	
Other (n = 46)	15	
Sex**		
Men (n = 2086)	10	NS
Women (n = 1767)	10	
Age**		
< 16 years (n = 396)	8	< 0.02
16-24 years (n = 189)	7	
25-44 years (n = 972)	9	
45-64 years (n = 880)	9	
≥ 65 years (n = 1453)	12	
Time from admission***		
≤ 24 hours (n = 634)	10	< 0.01
2 to 6 days (n = 1588)	8	
≥ 7 days (n = 1661)	11	

*   chi^2 test for prevalence rates
**   missing data
***   until stool culture

Characteristics of patients carrying C. difficile

Carriers of *C. difficile* were often affected by associated severe or chronic illness (52%), and/or undergoing treatment for other diseases: enteral feeding 13%, chemotherapy 13%, immunosuppressive therapy 9%, HIV positive status 11%.

Toxinogenicity and serotyping

The proportion of toxin producing strains was lower in the controls than in the cases, but the difference did not reach statistical significance. A toxigenic strain was more frequently isolated when the stool was loose, or when there was clinical diarrhea (Table 3). In contrast, the toxigenic strains were more often found in the absence of associated chronic or severe disease (Table 3).

The serotyping was carried out in 382 strains (371 cases and 11 controls). Four serotypes predominated: C, D, G and H. The distribution of the serotypes was different in the cases and controls. The proportion of strains belonging to group D was statistically higher in controls than in cases (45% vs. 18%, p = < 0.05). Conversely serotype C was isolated only in the cases.

Table 3. Prevalence of toxin producing-strains of *C. difficile*

Variables	Prevalence of toxigenic strains (%)*	p value
Consistency of stool**		
Diarrheal (n = 242)	72	< 0.001
Non diarrheal (n = 147)	55	
Clinical diarrhea**, ***		
Yes (n = 265)	71	< 0.001
No (n = 105)	51	
Chronic or debilitating underlying disease**		
Yes (n = 202)	60	< 0.01
No (n = 105)	74	
Antibiotic therapy		
Yes (n = 275)	65	NS
Non (n = 116)	68	

*   a toxigenic strain is defined by its ability to produce toxins A/B in vitro
** missing data
*** as noted by medical staff

Table 4. Characteristic of each serogroup in terms of toxin production, stool consistency, clinical diarrhea and antibiotic therapy

Variables			Serogroup				p value*
	C	D	G	H	Others	NT	
Toxigenic strain (%)	83	0	96	97	67	88	< 0.001
Loose stool (%)	83	46	74	67	69	58	< 0.01
Clinical diarrhea** (%)	96	57	81	76	80	66	< 0.01
Antibiotic therapy (%)	92	76	74	59	79	68	< 0.05

NT = non typable
*   chi² test among all serogroup types
** as noted by medical staff

In terms of characteristics of each serotype, the toxinogenicity appears to be associated with the serotype: group D was never found to be toxigenic; groups G and H were most frequently toxigenic. Group D was associated less often with diarrheal stools and with clinical diarrhea than the other groups. Group C was associated most with antibiotic treatment (Table 4).

Discussion

The aim of this multicentric study was to establish the prevalence and pathological significance of *C. difficile* when this organism is isolated from stools of hospital patients. For this, we compared the prevalence of *C. difficile* in the stools of hospital patients in whom a stool culture had been requested by the doctor and sent to the laboratory («cases»), with that of a control group of hospital patients. In these controls, a stool culture was performed by the laboratory although not requested by the doctor in charge of the patient.

It was thus assumed that the request for culture made by the doctor was indicative of gastrointestinal tract problems. Comparison of the records of the cases and controls validated the choice of the controls as they did not differ in the stratification criteria, namely referring team, age and length of hospital stay before stool collection. Nonetheless 19% of the controls had loose stools, without the doctor having considered these an indication for stool culture.

In this study, the search for *C. difficile* was standardized by the 11 laboratories using a commercial culture medium. Toxin B was estimated *in vitro* on culture of isolated bacterium (toxigenic culture) by a reference laboratory. These methods were chosen in the interests of standardization, as although the stool cytotoxicity test is a reference tool, it is not standardized. It has been shown that toxigenicity as measured on culture correlates well with the cytotoxicity test and has a high sensitivity [2].

The prevalence of *C. difficile* was twice as high in the cases as in the controls. Also, toxin producing strains were isolated more frequently from the cases than the controls, although the difference did not reach statistical significance, perhaps because of the low prevalence of *Clostridium difficile* carriers in the control group (11 patients).

Taking the appearance of the stools (as opposed to a request for stool culture) as an indicator of gastrointestinal tract pathology, the prevalence of *C. difficile* was three to four times as high in the case of diarrheal stools as in the normal stools of the controls. Also the toxin producing strains were more frequently isolated in the diarrheal stools and in the case of clinical diarrhea.

In this study, the prevalence of *C. difficile* in the cases was lower than that previously seen in patients with diarrhea and/or antibiotic induced colitis (15-25%). [1, 11]. This difference could be explained by: (1) the method of patient selection, which was not slanted in favor of those taking antibiotics; (2) the variability of the criteria used in defining diarrhea; (3) differing practices in requesting a stool culture (we have already indicated that 19% of the controls with loose stools were not thought by their doctors to require a stool culture).

The relation between the presence of *C. difficile* and other variables was studied by multivariate analysis using a logistic regression model. The age of the patients was independently related to the presence of *C. difficile*. This observation agrees with published data showing that advanced age and long stay, in geriatric or non geriatric units, increase the relative risk of acquiring *C. difficile* [1, 17].

We have also observed that the prevalence of *C. difficile* varies with the interval between the admission date and the request for stool culture. The prevalence was higher during the first 24 hours and after the second week. These differences although small were statistically significant. They suggest two separate ways of infection, namely nosocomial or community-acquired, for the patients in our study. Much previous work has shown that the risk of acquiring *C. difficile* rises with the length of hospital stay [4, 6, 10, 12, 14, 16, 19].

In our study, the lack of clinical data in respect of the patients who were not colonized by *C. difficile* precluded comparison with the carrier patients, so that it was not possible to establish the risk factors for acquiring the organism. The characteristics of the stool carriers of *C. difficile* (associated severe or chronic illness, treatment with chemotherapy or immunosuppressors, enteric feeding) agree with previous studies which define the risk factors [17].

The strains isolated from the cases and the controls were serotyped according to Delmée's method [5, 8]. This method, standardized by an ELISA technique, is reproducible, simple and rapid. Our study shows that four subgroups predominate in France: namely C, D, G, and H. As Delmée et al [7] have previously shown, serotype D has never been toxigenic. It was found mainly in the stools of the controls, and was statistically isolated less often from loose stools or from patients with clinical diarrhea. These results confirm the relation between toxinogenicity and serotype, and between pathogenicity and serotype.

In conclusion, our study has strengthened the relation between the presence of *C. difficile* in the stools of a hospital patient and the gastrointestinal pathology with which he/she is confronted. We recommend a routine search for *C. difficile* in any loose or liquid stool from a hospital patient.

References

1. Aronsson B, Möllby R, Nord CE (1985) Antimicrobial agents and *Clostridium difficile* in acute enteric disease: epidemiological data from Sweden, 1980-1982. J Infect Dis 151: 476-481
2. Barbut F, Kajzer C, Planas N, Petit JC (1993) Comparison of three enzyme immunoassays, a cytotoxicity assay, and toxigenic culture for the diagnosis of *Clostridium difficile*-associated diarrhea. J Clin Microbiol 31: 963-967
3. Brettle RP, Wallace E (1982) *Clostridium difficile* from stools of normal children. Lancet i, 1193
4 Clabots CR, Johnson S, Olson MM, Peterson LR, Gerding DL (1992) Acquisition of *Clostridium difficile* by hospitalized patients: evidence for colonized new admissions as a source of infection. J Infect Dis 166: 561-67
5. Delmée M, Homel M, Wauters G (1985) Serogrouping of *Clostridium difficile* by slide agglutination. J Clin Microbiol 21: 323-327
6. Delmée M, Vaudercam B, Avesani V, Michaux JL (1987) Epidemiology and prevention of *Clostridium difficile* infections in a leukemia unit. Eur J Clin Microbiol 6: 623-627
7. Delmée M, Avesani V (1990) Virulence of ten serogroups of *Clostridium difficile* in hamsters. J Med Microbiol 33: 85-90
8. Delmée M, Depitre C, Corthier G, Ahoyo A, Avesani V (1993) Use of enzyme-linked immunoassay for *Clostridium difficile* serogrouping. J Clin Microbiol 31: 2526-2528
9. Delmée M, Buts JP (1993) *Clostridium difficile*-associated diarrhoea in children. In: Buts JP, Sokal EM (eds) Management of digestive and liver disorders in infants and children. Elsevier, Amsterdam, pp 371-379
10. Fekety JP, Kim KH, Brown D, et al (1981) Epidemiology of antibiotic-associated colitis: isolation of *Clostridium difficile* from the hospital environment. Am J Med 70: 906

11. George WL, Rolfe RD, Finegold SM (1982) *Clostridium difficile* and its cytotoxin in feces of patients with antimicrobial agent-associated pseudomembranous colitis and miscellaneous conditions. J Clin Microbiol 15: 1049

12. Heard SR, O'Farrell S, Holland D, Crook S, Barnett MJ, Tabaqchali S (1986) The epidemiology of *Clostridium difficile* with use of a typing scheme: nosocomial acquisition and cross-infection among immunocompromised patients. J Infect Dis 153: 159-162

13. Holst E, Helin I, Per-Anders M (1981) Recovery of *Clostridium difficile* from infants. Scand J Infect Dis 13: 41-45

14. Johnson S, Clabots CR, Linn FV, Olson MM, Peterson LR, Gerding DN (1990) Nosocomial *Clostridium difficile* colonisation and disease. Lancet 336: 97-100

15. Kelly CP, Pothoulakis C, LaMont JT (1994) *Clostridium difficile* colitis. N Engl J Med 330: 257-62

16. McFarland LV, Mulligan ME, Kwok RYY, Stamm WE (1989) Nosocomial acquisition of *Clostridium difficile* infection. N Engl J Med 320: 204-10

17. McFarland LV, Surawicz CM, Stamm WE (1990) Risk factors for *Clostridium difficile* carriage and *Clostridium difficile*-associated diarrhea in a cohort of hospitalized patients. J Infect Dis 162: 678-684

18. McFarland LV, Elmer GW, Stamm WE, Mulligan ME (1991) Correlation of immunoblot type, enterotoxin production, and cytotoxin production with clinical manifestations of *Clostridium difficile* infection in a cohort of hospitalized patients. Infect Immun 59: 2456-2462

19. Samore MH, DeGirolami PC, Tlucko A, Lichtenberg DA, Melvin ZA, Karchmer AW (1994) *Clostridium difficile* colonization and diarrhea at a tertiary care hospital. Clin Infect Dis 18: 181-87

20. Viscidi R, Willey S, Bartlett JG (1981) Isolation rates and toxigenic potential of *Clostridium difficile* isolated from various patient populations. Gastroenterology 81: 5-9

Epidemiological markers of *Clostridium difficile*

Soad Tabaqchali, Mark Wilks

SUMMARY

The identification of *Clostridium difficile* as a major enteric pathogen that causes nosocomial infections has necessitated the development of typing methods both to provide a better understanding of the epidemiology of this organism, and some insight into the pathogenicity of various strains. Several fingerprinting and typing methods have been developed based on phenotypic and molecular epidemiological markers. Phenotypic markers include antibiotic resistance, bacteriocin and bacteriophage susceptibility patterns and protein electrophoretic profiles including radiolabelled proteins. Immunological markers have been used to develop typing methods based on Western blotting profiles and serogrouping using slide agglutination. More recently the profiles of volatile products produced by pyrolysis mass spectrometry (PMS) have also been used in the fingerprinting of *C. difficile*.

Genotypic markers such as plasmid profiles, DNA restriction endonuclease and ribosomal rRNA restriction patterns have also been used in fingerprinting *C. difficile*. These have the advantage of stability, but are often more time consuming to perform than the phenotypic methods. However, serogrouping and phage typing, although simple to apply, require the maintenance of stocks of antisera and phages. More recently the application of PCR with arbitrary primers to amplify the rRNA intergenic spacer regions have been successfully applied in the fingerprinting of *C. difficile* strains.

Despite the avalanche of techniques, there is still no interlaboratory standardisation of methods and no unified nomenclature. Attempts are currently being undertaken to exchange type strains and establish an international nomenclature.

Marqueurs épidémiologiques de *Clostridium difficile*

RÉSUMÉ

L'identification de *C. difficile* comme germe entéropathogène responsable d'infections nosocomiales a nécessité le développement de méthodes de typage, pour permettre une meilleure compréhension de l'épidémiologie de ce microorganisme, ainsi qu'une connaissance plus approfondie de la pathogénicité des différentes souches.

Plusieurs techniques d'empreinte génétique et de typage ont été développées à partir des marqueurs épidémiologiques (marqueurs phénotypiques ou génotypiques). Les marqueurs phénotypiques comprennent la résistance aux antibiotiques, les profils de sensibilité aux bactériocines (bactériocinotypie), aux bactériophages (lysotypie), et les profils de migration électrophorétique des protéines par marquage radioactif. Des marqueurs immunologiques ont été utilisés pour développer des méthodes de typage basées sur la mise en évidence des protéines spécifiques par "Western blot", ou sur le sérotypage par la technique d'agglutination sur lame. Plus récemment, les profils chromatographiques des acides gras volatils obtenus par spectrométrie de masse après pyrolyse (PMS) ont également été utilisés pour identifier *C. difficile*.

Les marqueurs génotypiques tels que les profils plasmidiques, les profils de restriction des ADN en général et en particulier de l'ADN ribosomal ont contribué à établir une empreinte génétique de *C. difficile*. Ces techniques présentent l'avantage d'être reproductibles, mais elles sont plus longues à réaliser que les méthodes phénotypiques. Cependant, le sérogroupage et le typage par phages, bien que simples à exécuter, demandent le maintien de stocks d'antisérums et de phages. Plus récemment, l'application de la PCR (Polymerase Chain Reaction) avec utilisation des amorces aléatoires, ainsi que la technique d'amplification d'une séquence cible située dans la région non codante de l'ARN ribosomal ont été appliquées avec succès pour l'établissement d'empreintes génétiques de différentes souches de *C. difficile*.

En dépit de la multiplication des techniques, il n'y a pas encore de standardisation commune à l'ensemble des laboratoires, et pas de nomenclature unique. Un objectif général est d'entreprendre l'échange des différentes souches afin d'établir une nomenclature internationale.

Epidemiologische Marker für *Clostridium difficile*

ZUSAMMENFASSUNG

Die Identifikation von *C. difficile* als wichtigstes Darmpathogen und Auslöser nosokomialer Infektionen machte die Entwicklung von Typisierungsmethoden notwendig, um die Epidemiologie dieses Keims besser zu verstehen und zugleich Einblicke in die Pathogenität bestimmter Stämme zu gewinnen. Mehrere Fingerprinting- und Typisierungsmethoden wurden ausgehend von phänotypischen und molekularen epidemiologischen Markern entwickelt. Zu den phänotypischen Markern gehören Antibiotikaresistenz, Empfindlichkeitsmuster gegenüber Bakteriocinen und Bakteriophagen sowie Eiweißelektrophorese-Profile einschließlich radioaktiv markierter Proteine. Immunologische Marker werden zur Entwicklung von Typisierungsverfahren ausgehend vom Western Blotting Test und der Bestimmung serologischer Gruppen mittels Objektträgeragglutination eingesetzt. Seit kurzem werden auch die Profile der bei der pyrolytischen Massenspektrometrie (PMS) entstehenden flüchtigen Produkte für das Fingerprinting von *C. difficile* eingesetzt.

Genotypische Marker wie Plasmidprofile, DNA-Restriktionsendonuklease- und rRNA-Restriktionsmuster wurden ebenfalls für das Fingerprinting von *C. difficile* herangezogen. Sie haben den Vorteil, stabil zu sein, sind allerdings oft zeitaufwendiger als phänotypische Methoden. Die Bestimmung serologischer Gruppen und die Phagentypisierung sind zwar einfach anzuwendende Methoden, erfordern jedoch einen Vorrat an Antiseren und Phagen. Neuerdings werden im Rahmen des Fingerprinting von *C. difficile*-Stämmen Polymerase-Kettenreaktionen mit Zufallsprimern oder zur Vermehrung von intergenen Spacersequenzen der rRNA erfolgreich eingesetzt.

Trotz der Vielfalt an Verfahren steht eine Standardisierung der Methoden zwischen den verschiedenen Labors noch ebenso aus wie eine einheitliche Nomenklatur. Gegenwärtig sind Versuche im Gange, Stammtypen auszutauschen und eine internationale Nomenklatur zu erstellen.

Introduction

Clostridium difficile is the commonest nosocomial enteric pathogen. It causes pseudomembranous colitis and is strongly associated with antibiotic-associated diarrhoea and colitis in man [1, 11].

The recognition of the importance of this organism and the ever increasing worldwide reports of outbreaks of *C. difficile*-associated disease (CDAD) in hospitals and institutions has necessitated the development of typing schemes to study the epidemiology of this organism and to provide some insight into the pathogenicity of various strains.

Over the past decade, numerous typing and fingerprinting methods have been developed based on various epidemiological markers of *C. difficile* and have been applied in epidemiological studies [27]. These studies have demonstrated clearly that nosocomial acquisition of *C. difficile* and cross-infection among patients occur and that outbreaks are usually caused by a single type-strain [13, 15, 19].

This paper presents a brief outline of the typing and fingerprinting methods based on phenotypic or genotypic markers with an assessment of their suitability for use as typing methods for *C. difficile*.

Phenotypic markers

Toxin production

Toxin production by strains of *C. difficile* is an important marker but of limited value since only the presence or the absence of toxin is measured.

Antibiotic susceptibility patterns

Antibiograms may occasionally differentiate strains but they are of limited value because the MICs of *C. difficile* strains fall within a narrow range and most strains are either sensitive or resistant. Antibiograms however, may be helpful in combination with other fingerprinting or typing methods [9].

Bacteriophage/Bacteriocin susceptibility patterns

The susceptibility of *C. difficile* strains to various sets of bacteriocins and bacteriophages has been utilised to differentiate strains by Sell et al [25]. This method was used to type 114 *C. difficile* strains producing 31 typing patterns [18]. However, only 16-40% of strains were typable. Furthermore, the set of phages is not widely available.

Electrophoretic protein patterns

Staining with Coomassie blue

The separation of cellular proteins of *C. difficile* by SDS-polyacrylamide gel electrophoresis (PAGE), followed by staining with Coomassie blue produces distinct protein profiles allowing the detection of similarities of pattern during an outbreak. This is a simple method which can be applied in most laboratories, but standardisation of procedures is necessary to obtain reproducible results.

[35]S-methionine labelled protein profiles (Radio-PAGE)

This method is based on the incorporation of ^{35}S-methionine into cellular proteins, their separation on SDS-PAGE and subsequent autoradiography [28]. The method has enabled the recognition of 15 distinct types of C. difficile [29], and more recently the scheme has been extended to 20 types (Al-Saleh and Tabaqchali, unpublished data]. The types are based on the presence of specific and consistent radiolabelled major bands. This technique has been applied extensively in epidemiological studies. The radio-PAGE strain types have been confirmed to be distinct by immunoblotting [14], restriction endonuclease analysis [31] and PCR typing [30].

Immunological markers

Immunoblotting

Hyperimmune antisera raised in rabbits against C. difficile strains have been used in Western blotting to differentiate between strains of C. difficile [14, 22]. A single antiserum is inadequate to discriminate amongst strains. The presence of shared common antigens was demonstrated amongst nine standard radio-PAGE types when immunoblotted against heterologous and homologous antisera raised against the nine types, and the type-specific response was only present with homologous antisera [14]. Mulligan et al [22] found that two antisera recognised ten immunoblot types and that immunoblotting was more discriminatory than PAGE and serogrouping.

Serogrouping

Antisera raised against *Clostridium difficile* strains were used in slide agglutination of formol-treated cells. Ten serogroups of C. difficile have been identified following cross-absorption of the sera, and these were designated A–D, F–I, K and X [6]. Some serogroups were homogeneous but others were heterogenous, particularly serogroup A which contained at least 12 different electrophoretic patterns within the group [7]. The use of deflagellated cells of C. difficile in slide agglutination reduced some cross reactions. Serogrouping has been applied extensively in epidemiological studies [8]. The method is simple to perform but does depend on the availability of banks of sera. Antisera are not available commercially, and the production and quality control of a set of typing sera would be extremely expensive.

Pyrolysis mass spectrometry (PMS)

Thermal degradation of bacterial cells produce a wide range of volatile products, the pyrolysate, which reflect the organic composition of the cells. The pyrolysate is then passed through a mass spectrometer and multivariant analysis performed using specialised computer software [10]. This technique has been applied to C. difficile strains isolated from two outbreaks of CDAD in hospitals. Clustering of patterns was demonstrated indicating a single outbreak strain was involved in each hospital [26]. PMS however, remains in the experimental stage and requires validation of reproducibility of results. It is useful as a fingerprinting but not currently as a typing method. Although the running costs are low, its very high capital cost (£50,000) prohibits its general applicability.

Genotypic markers

Epidemiological markers based on genotypic characteristics are generally more discriminating and specific.

Plasmid analysis

The percentage of isolates of *C. difficile* which contain plasmids varies from 18% [21] to 59% [18], thus the detection of plasmids may only be useful in follow up analysis of strains harbouring specific plasmid patterns during an outbreak [5]. Analysis of plasmids has been used as an epidemiological marker for *C. difficile* in conjunction with other fingerprinting and typing methods [32].

Restriction endonuclease analysis (REA)

REA of genomic DNA has been used by several investigators for fingerprinting of *C. difficile* strains [17, 24, 31]. This technique provides numerous definitive and discriminating bands but the resultant complex pattern is difficult to analyse. Characterisation of strains of *C. difficile* by REA has shown correlation with radio-PAGE types [31]. REA of nine radio-PAGE types gave distinct individual patterns and analysis of type X strains isolated from 10 different patients during an outbreak of AAC revealed indistinguishable *Hin* d III DNA profiles, confirming that a single strain caused the outbreak [31].

Ribotyping

Ribosomal rRNA from *Escherichia coli* has been used as a probe to detect restriction length fragment polymorphisms (RFLPs) using enhanced chemiluminescence within *C. difficile* digested genomic DNA. The ribotyping pattern which has fewer bands, is simpler and easier to interpret than that produced by REA [2]. However, it lacks resolution and requires an additional step of hybridisation which makes it more time consuming and labour intensive.

Polymerase chain reaction with arbitrary primers AP-PCR

The application of this technique was first described by McMillin and Muldrow [20]. Extracted DNA from *C. difficile* strains was amplified with a panel of six 10bp arbitrary primers.

One random primer was used in each reaction and the resultant patterns were compared. These provide different AP-PCR profiles and it is suggested that in combination with an initial PCR using primers for toxins A or B, the method may prove useful in the rapid differentiation of toxigenic and non-toxigenic strains and subsequent fingerprinting.

The AP-PCR method was further simplified by Wilks and Tabaqchali [30] by applying it directly to colonies of *C. difficile* grown on blood agar plates without the necessity of DNA extraction. A single arbitrary primer was used and applied to the nine radio-PAGE types described above producing specific patterns. When applied to outbreak strains previously shown to be of a single radio-PAGE type, a single pattern for the outbreak strain was demonstrated. AP-PCR has also been applied to group 41 isolates into nine groups [16].

PCR of rRNA intergenic spacer regions

The rRNA operon has some regions which are highly conserved and others that are variable. The length of the 16S rRNA gene is constant but there are variable length 16S-23S rRNA spacer regions [23]. Sixteen different rRNA alleles have been described, with a single isolate of *C. difficile* possessing between five and nine different alleles. PCR across this region is used to amplify the rDNA to produce the different profiles among *C. difficile* strains [12].

PCR-ribotyping was used to investigate an outbreak of *C. difficile*-associated diarrhoea and compared with REA of genomic DNA. Both methods showed that five out of six patients were infected with a single strain [4].

In our laboratory, this method was applied successfully directly to colonies of *C. difficile*, eliminating the need for the cumbersome DNA extraction procedure. rRNA-PCR was applied to 20 radio-PAGE type strains and each showed a specific profile (see Fig. 1). Furthermore, strains isolated during an outbreak of CDAD produced indistinguishable rRNA-PCR types [unpublished data]. PCR ribotyping provides a possible typing method for *C. difficile* based on the variable length of 16S-23S rRNA spacer regions. It is simple, rapid and reproducible and may provide a reliable method for epidemiological studies.

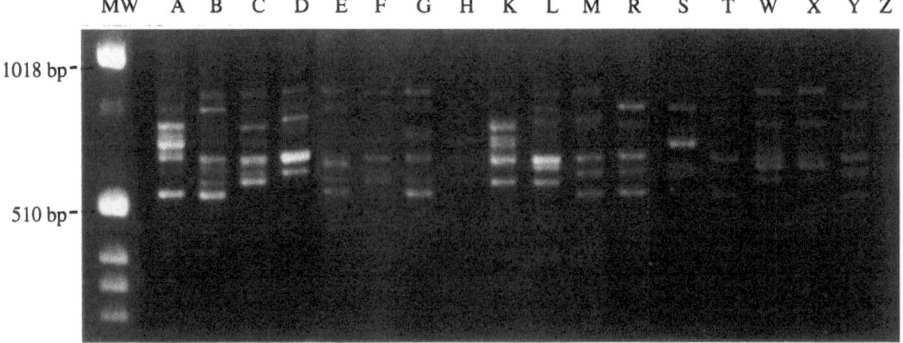

Fig. 1. PCR ribotyping of radio-PAGE type strains of *C. difficile*. Bacterial colonies were tested directly without prior extraction of DNA. Primers and PCR conditions were those of Cartwright et al [4]

Conclusions

The identification of several phenotypic and genotypic epidemiological markers of *C. difficile* has facilitated the development of fingerprinting and typing methods for this organism as outlined in this paper. Most of these are fingerprinting methods, useful as comparative techniques identifying clusters of strains during an outbreak.

Only a few techniques have attempted to provide typing schemes with well characterised distinct type strains which have been tested by several methods. The new genotypic fingerprinting methods have the advantage of stability but they are often more time consuming to perform than the phenotypic methods. The introduction of PCR technology may provide a simpler and more rapid method. Although the use of PCR with arbitrary primers is rapid, giving a same day result, amplification of consistent DNA fingerprints with short arbitrary primers is difficult and is unlikely to yield reproducible results between laboratories. PCR-ribotyping using specific primers appears to be potentially a more valuable technique.

There have been very few attempts at interlaboratory standardisation of techniques and exchange of type strains. Unless this approach is accepted, then the above methods will remain limited to their institutions. An initial approach towards standardisation by producing a unified nomenclature for *C. difficile* type strains has been proposed [3]. This includes country of origin, major type by method, subtype and toxigenicity. For example, the toxigenic epidemic strain (type X) at St. Bartholomew's Hospital, London, UK would be designated: UK/radio-PAGE/type X/A⁺B⁺. Exchange of strains which could then be tested by other methods to provide a unified nomenclature and universal standard types will be an invaluable contribution to our understanding of the global epidemiology of this important nosocomial pathogen.

Acknowledgements
We wish to acknowledge the help of Ms. Sandy Gale in the preparation this manuscript.

References

1. Bartlett JG, Chang TW, Gurwith M, Gorbach SL, Onderdonk AB (1978). Antibiotic-associated pseudomembranous colitis due to toxin-producing clostridia. N Engl J Med 298: 531-534
2. Bowman RA, O'Neill GL, Riley TV (1991) Non-radioactive restriction fragment length polymorphism (RFLP) typing of *Clostridium difficile*. FEMS Microbiol Lett 63: 269-272
3. Brazier JS, Delmée M, Tabaqchali S, Hill LR, Mulligan ME, Riley TV (1994) Proposed unified nomenclature for *Clostridium difficile* typing. Lancet 343: 1578-1579
4. Cartwright CP, Stock F, Beckmann SE, Williams EC, Gill VJ (1995) PCR amplification of rRNA intergenic spacer regions as a method for epidemiologic typing of *Clostridium difficile*. J Clin Microbiol 33: 184-187

5   Clabots C, Lee S, Gerdin D, Mulligan M, Kwok R, Schaberg D, et al (1988) *Clostridium difficile* plasmid isolation as an epidemiological tool. Eur J Clin Microbiol Infect Dis 7: 312-315

6. Delmée M, Homel M, Wauters G (1985) Serogrouping of *Clostridium difficile* strains by slide agglutination. J Clin Microbiol 21: 232-237

7. Delmée M, Laroche V, Avesani V, Cornelis G (1986) Comparison of serogrouping and polyacrylamide gel electrophoresis for typing *Clostridium difficile*. J Clin Microbiol 24: 991-994

8. Delmée M, Bulliard G, Simon G (1986) Application of a technique for serogrouping *Clostridium difficile* in an outbreak of antibiotic-associated diarrhoea. J Infect 13: 5-9

9. Delmée M, Avesani V (1988) Correlation between serogroup and susceptibility to chloramphenicol, clindamycin, erythromycin, rifampicin and tetracycline among 308 isolates of *Clostridium difficile*. J Antimicrob Chemother 22: 325-331

10. Freeman R, Goodfellow M, Gould FK, Hudson SJ, Lightfoot NF (1990) Pyrolysis mass spectrometry (Py-MS) for rapid epidemiological typing of clinically significant bacterial pathogens. J Med Microbiol 32: 283-286

11. George RHS, Symonds JM, Dimock F, Brown JD, Arabi Y, Sinagawa N, Keighley MRB, Alexander-Williams J, Burdon DW (1978) Identification of *Clostridium difficile* as a cause of pseudomembranous colitis. Br Med J 1: 695

12. Gurtler V (1993) Typing of *Clostridium difficile* strains by PCR-amplification of variable length 16S-23S rDNA spacer regions. J Gen Microbiol 139: 3089-3097

13. Heard S, O'Farrell S, Holland D, Crook S, Barnett MJ, Tabaqchali S (1986) The epidemiology of *Clostridium difficile* with the use of a typing scheme: nosocomial acquisition and cross-infection among immunocompromised patients. J Infect Dis 153: 159-162

14. Heard SR, Rasburn B, Matthews RC, Tabaqchali S (1986) Immunoblotting to demonstrate antigenic immunogenic differences among nine standard strains of *Clostridium difficile*. J Clin Microbiol 24: 384-387

15. Johnson S, Clabots CR, Linn FV, Olson MM, Peterson LR, Gerding DN (1990) Nosocomial *Clostridium difficile* colonisation and disease. Lancet 336: 97-100

16. Killgore FE, Kato H (1994) Use of arbitrary primer PCR to type *Clostridium difficile* and comparison with those by immunoblot typing. J Clin Microbiol 32: 1591-1593

17. Kuijper EJ, Oudbuer JH, Shufbergen W, Jansz A, Zanen HC (1987) Application of whole cell DNA restriction endonuclease profiles to the epidemiology of *Clostridium difficile* induced diarrhoea. J Clin Microbiol 25: 751-753

18. Mahony DE, Clow J, Atkinson L, Vakharia N, Schlech WF (1991) Development and application of a multiple typing system for *Clostridium difficile*. Appl Environ Microbiol 57: 1873-1879

19. McFarland LV, Mulligan ME, Kwok RYY, Stamm WE (1989) Nosocomial acquisition of *Clostridium difficile* infection. N Engl J Med 320: 204-210

20. McMillin DE, Muldrow LL (1992) Typing of toxic strains of *Clostridium difficile* using DNA fingerprints generated with arbitrary polymerase reaction primers. FEMS Microbiol Lett 92: 5-19

21. Muldrow LL, Archibold ER, Nunez-Montiel OL, Sheehy RJ (1982) Survey of the extra-chromosomal gene pool of *Clostridium difficile*. J Clin Microbiol 16: 637-640

22. Mulligan NE, Peterson LR, Kwok RYY, Clabots LR, Gerding DN (1988) Immuno-blots and plasmid fingerprints compared with serotyping and polyacrylamide gel electrophoresis for typing *Clostridium difficile*. J Clin Microbiol 26: 41-46

23. Neefs J-M, van de Peer Y, Hendriks L, de Wachter R (1990) Compilation of small ribosomal sub-unit RNA sequences. Nucleic Acids Res 18 (suppl): S2237-2317

24. Peerbooms PGH, Kuijt P, MacLaren DM (1987) Application of chromosomal restriction endonuclease digest analysis for use as a typing method for *Clostridium difficile*. J Clin Pathol 40: 771-776

25. Sell TL, Schaberg DR, Fekety FR (1983) Bacteriophage and bacteriocin typing scheme for *Clostridium difficile*. J Clin Microbiol 17: 1148-1152

26. Sisson PR, Freeman R, Lightfoot NF (1993) Outbreaks of *Clostridium difficile* infection investigated by pyrolysis mass spectrometry. PHLS Microbiology Digest 10: 100-101

27. Tabaqchali S, Wilks M (1992) Epidemiological aspects of infections caused by Bacteroides fragilis and *Clostridium difficile*. Eur J Clin Microbiol Infect Dis 11: 1049-1057

28. Tabaqchali S, Holland D, O'Farrell S, Hilman R (1984) Typing scheme for *Clostridium difficile*: Its application in clinical and epidemiological studies. Lancet i: 935-938

29. Tabaqchali S (1990) Epidemiologic markers of *Clostridium difficile*. Rev Infect Dis 12: S192-S199

30. Wilks M, Tabaqchali S (1994) Typing of *Clostridium difficile* by polymerase chain reaction with an arbitrary primer. J Hosp Infect 28: 231-234

31. Wren M, Tabaqchali S (1987) Restriction endonuclease DNA analysis of *Clostridium difficile*. J Clin Microbiol 25: 2402-2403

32. Wust J, Sullivan NM, Hardegger U, Wilkins TD (1982) Investigation of an outbreak of antibiotic-associated colitis by various typing methods. J Clin Microbiol 16: 1096-1101

Recent advances in the structure and function of *Clostridium difficile* toxins

J. Thomas LaMont

SUMMARY

Toxins A and B from *Clostridium difficile* mediate the tissue damage, diarrhea and intestinal inflammation which occur during infection with this pathogen. The toxins are encoded by two genes *tox* A and *tox* B which lie in close proximity on the bacterial chromosome. Nearly all pathogenic strains produce both toxins, while non-toxigenic strains produce neither. The Mr of toxin A is 308,057 with a pI of 5.3. Neither toxins are glycosylated, and are denatured by heat and at pH above 8.0 or below 5.0. The toxins have considerable amino acid homology and nearly identical spacing of cysteine residues suggesting that their secondary structures are similar. The carboxyl-terminal third of both toxins consists of repeating peptides of 20 to 30 amino acids with extensive homology. These repeats contain the receptor binding domain by which the toxin attaches to the plasma membrane of target cells. The catalytic or (enzymatic) activity of toxins A and B is identical; both toxins are specific monoglucosyltransferases that catalyze glucose transfer from UDP-glucose to threonine in position 37 of *rho* A. *Rho* A is a low molecular weight GTP-binding protein of the *ras* superfamily and is a major regulator of actin filament formation in cells. This enzymatic modification of *rho* proteins by *C. difficile* toxins explains the disaggregation of actin filaments in exposed cells.

The intestinal effects of toxins are species specific in that toxin B is approximately 10 times more potent than toxin A in human colon, while in rodent intestine only toxin A produces inflammation or diarrhea. These differences probably relate to receptor differences between rodents and humans. Luminal exposure of human or rodent colon or ileum to toxins causes actin filament disaggregation in enterocytes resulting in increased paracellular permeability. This is followed by an acute inflammatory reaction in the lamina propria that involves activation of mast cells, upregulation of leukocyte adhesion molecules on vascular endothelium, release of the neuropeptide substance P and other cytokines and chemokines. The net result is infiltration of the bowel wall with neutrophils and pseudomembranous colitis with exudative diarrhea.

Nouvelles connaissances sur la structure et la fonction des toxines de *Clostridium difficile*

RÉSUMÉ

Les toxines A et B de *C. difficile* sont responsables des lésions tissulaires, de la diarrhée et des inflammations intestinales survenant au cours d'une l'infection par ce germe pathogène. Les toxines sont codées par deux gènes, *tox* A et *tox* B, qui sont situés à proximité sur le chromosome bactérien. Pratiquement toutes les souches pathogènes produisent les deux toxines, alors que les souches non toxinogènes n'en produisent aucune. Le poids moléculaire relatif de la toxine A est de 308.057, avec un point isoélectrique égal à 5,3. Aucune toxine n'est glycosylée, et elles sont toutes deux dénaturées par la chaleur ou par des pH supérieurs à 8 ou inférieurs à 5. Les toxines présentent des analogies considérables quant aux acides aminés, et l'espacement pratiquement identique des résidus cystéine suggèrent que leurs structures secondaires sont similaires. Le dernier tiers du côté C terminal des deux toxines est formé de séquences peptidiques répétitives, constituées de 20 à 30 acides aminés qui présentent une grande homologie entre eux. Dans ces séquences répétées, on trouve le site de liaison de la toxine par lequel elle se fixe à la membrane plasmique des cellules cibles. L'activité catalytique ou enzymatique des toxines A et B est identique. Les deux toxines possèdent une activité monoglucosyltransférase spécifique, qui catalyse le transfert du glucose de l'UDP-glucose à la thréonine en position 37 de *rho* A. *Rho* A est une protéine de faible poids moléculaire, et dont la fixation est GTP-dépendante. Elle appartient à la grande famille des protéines *ras*, et joue un rôle de régulateur important dans la formation des filaments d'actine dans les cellules. C'est la modification enzymatique des protéines *rho* par les toxines de *C. difficile* qui explique la désagrégation des filaments d'actine dans les cellules exposées.

Les effets intestinaux des toxines sont spécifiques de l'espèce considérée. Chez l'homme, la toxine B est environ 10 fois plus puissante que la toxine A au niveau du côlon, alors que dans l'intestin du rongeur, seule la toxine A produit une inflammation ou une diarrhée. Ces différences reflètent probablement celles qui existent au niveau des récepteurs entre les rongeurs et les humains. Chez l'homme ou le rongeur, l'exposition intraluminale du côlon ou de l'iléon aux toxines entraîne une désagrégation des filaments d'actine dans les entérocytes, ce qui provoque une augmentation de la perméabilité paracellulaire. Cela est suivi d'une réaction inflammatoire aiguë de la lamina propria, qui provoque l'activation des mastocytes, une surexpression des molécules d'adhésion des leucocytes sur l'endothélium vasculaire, la libération de substance P, d'autres cytokines et de chemokines. Le résultat final est l'infiltration de la paroi intestinale par des neutrophiles, et la présence d'une colite pseudomembraneuse avec diarrhée exsudative.

Jüngste Fortschritte in der Erforschung von Struktur und Funktion der *Clostridium difficile*-Toxine

ZUSAMMENFASSUNG

Die *C. difficile*-Toxine A und B bedingen die bei einer Infektion mit diesem pathogenen Keim auftretenden Gewebeschäden, Diarrhöen und Darmentzündungen. Die Toxine werden durch die beiden Gene *tox* A und *tox* B kodiert, die nah beieinander auf dem Chromosom des Bakteriums liegen. Fast alle pathogenen Stämme produzieren beide Toxine, nicht toxinogene Stämme dagegen keines von beiden. Das Mr des Toxins A beträgt 308,057 bei einem pI von 5,3. Beide Toxine sind glykosyliert, sie werden durch Hitze sowie bei einen pH-Wert von über 8,0 oder unter 5,0 denaturiert. Die Toxine weisen eine beträchtliche Aminosäurenhomologie und praktisch identische Abstände zwischen den Cysteinresten auf, was darauf schließen läßt, daß ihre sekundären Strukturen ähnlich sind. Das Carboxyl-terminale Drittel besteht bei beiden Toxinen aus sich wiederholenden Peptiden aus 20 bis 30 Aminosäuren mit hoher Homologie. Diese Sequenzen enthalten die Rezeptorbindungsdomäne, mit der das Toxin an die Plasmamembran der Zielzelle bindet. Die katalytische (oder enzymatische) Aktivität der Toxine A und B ist identisch; beide Toxine sind spezifische Monoglucosyltransferasen, die den Transfer von UDP-Glukose zum Threonin an Position 37 von *rho* A katalysieren. *Rho* A ist ein niedermolekulares GPT-bindendes Eiweiß der *ras*-Superfamilie und ein wichtiger Regulator der intrazellulären Bildung von Aktinfilamenten. Diese enzymatische

Modifikation von *rho*-Proteinen durch *C. difficile*-Toxine erklärt den Zerfall von Aktinfilamenten in den betroffenen Zellen.

Die intestinalen Wirkungen der Toxine sind artspezifisch insofern, als Toxin B im menschlichen Dickdarm etwa 10mal wirksamer ist als Toxin A, während bei Nagern nur Toxin A Darmentzündungen oder Diarrhöen auslöst. Diese Unterschiede beruhen vermutlich auf den bei Menschen und Nagern unterschiedlichen Rezeptoren. Bringt man die Toxine ins Lumen von Dick-oder Dünndarm von Menschen bzw. Nagern ein, kommt es zu einem Zerfall der Aktinfilamente in den Darmzellen und infolgedessen zu einer erhöhten parazellulären Permeabilität. Die Folge ist eine akute entzündliche Reaktion der Lamina propria mit Mastzellenaktivierung, Vermehrung der leukozytären Adhäsionsmoleküle am vaskulären Endothel, Freisetzung das Neuropeptids Substanz P sowie weiterer Zytokine und Chemokine. Das Ergebnis ist letzten Endes die Invasion von Neutrophilen in die Darmwand und eine pseudomembranöse Kolitis mit exsudativer Diarrhö.

Introduction

The toxins of *Clostridium difficile*, designated toxins A and B in the scientific literature, are central to the pathophysiology of diarrhea and colitis in experimental animal models and in man. Only toxigenic strains of the organism produce disease, and neutralization of toxins by specific antitoxin antibody prevents disease, at least in animals. The two toxins share many biologic properties, related to their very similar genetic and amino acid sequences. In human colonic tissues, both toxins cause necrosis and apoptosis of surface epithelial cells while in rodent models only toxin A possesses enterotoxic activity. These species differences between the toxins probably reflect different receptors for each toxin, as recent studies indicate that their intracellular mechanisms are identical. This brief update will focus on recent advances regarding *C. difficile* toxin genes, biologic activities in target cells and tissues, receptor binding studies and intracellular mechanisms of action.

Toxin genes

Genes encoding *Clostridium difficile* toxins A and B have been cloned and sequenced (Table 1). The toxins are encoded by two separate genes which lie in close proximity on the bacterial chromosome; *tox* B resides about 1 kilobase upstream of *tox* A [10]. The *tox* A and *tox* B genes are closely related structurally and biologically. Amino acid sequence analysis reveals an exact match of 44.8%

Table 1. *Clostridium difficile* genes and toxins

	Toxin A	Toxin B
Gene length	8130	7098
Deduced Mr of protein	308, 057	269, 709
pI of protein	5.3	4.1
Peptide repeats in C-terminus	30	19
Amino acid sequence homology	63%	63%
Enzymatic activity	Glucosyltransferase	Glucosyltransferase
Substrate in target cells	Rho A	Rho A

and 63.1% for exact plus conserved amino acids. The location of cysteines in the peptide chains is very similar in both toxins indicating similarity of secondary structure. The coding region of *tox* A contains 8130 base pairs and encodes a polypeptide of predicted Mr 308,057 with a pI of 5.3 [6]. The *tox* B gene is 7098 base pairs in length and encodes a protein of predicted Mr 269,709 with a pI of 4.1. Between the two toxin genes is an open reading frame, identified as *utx* A; another open reading frame, called *dtx* A, resides downstream from the *tox* A gene. These four open reading frames *tox* B, *utx* A, *tox* A and *dtx* A are encoded by a 29-kilobase fragment of *C. difficile* DNA.

A major structural feature of both toxin genes is the presence of SRONs (short repetitive oligonucleotide sequences) at the 3' end of the gene [6]. SRONs comprise approximately one third of the DNA sequence, and encode a series of repeating peptides or CROPs (combined repetitive oligopeptides) that make up the carboxy terminal one third of the molecule. These peptides in toxin A have been classified as class I (30 amino acids) or class II (20 or 21 amino acids) with extensive homology in both types of repeats. The repeating units in the peptide are highly immunogenic, and mediate toxin-receptor binding. The C-terminal repeats of toxin A bear striking sequence homology with glucosyltransferase enzymes of *Streptococcus mutans* and other Streptococci [24]. These streptococcal enzymes and the toxins of *C. difficile* share similar structure-function relationships. The carboxyl terminal third of the proteins consist of tandem repeats that bind ligands, while the amino terminal domain possesses the enzymatic or catalytic portion. Recent evidence suggest that toxins A and B are glucosyltransferase enzymes that inactivate small GTP-binding proteins within target cells (Table 1). This enzymatic activity resides in the non-repeating portions of the toxin molecule that constitutes the amino terminal two-thirds of the peptide chain.

Biological activities of toxins

As might be deduced from their striking genetic similarities, toxins A and B of *C. difficile* share a number of common biologic properties. The actions of these

Table 2. Biologic actions of toxins A and B

Action on cells or tissue	Tox A	Tox B	Comment
Hemagglutination of rabbit RBCs	+	-	High toxin concentrations
Mouse lethality	+	+	B more potent than A
Fibroblast rounding	+	+	B more potent than A
Neutrophil chemoattraction	+	-	High concentrations only
Enterotoxicity			
Human colon (in vitro)	+	+	B more potent than A
Rodent intestine (in vivo)	+	-	B effective only if bowel initially damaged

toxins on target cells and tissues are summarized in Table 2. The toxins are both lethal when injected into animals. For example, injection of infant rhesus monkeys with 6 to 11 µg of purified toxin A caused death within 3 to 8 hours [1]. Animals became lethargic and hypothermic and died of terminal apnea, preceded by convulsions and limb paralysis. Autopsy findings included visceral congestion and hemorrhage, but specific organ necrosis was absent. The observed findings are consistent with skeletal and cardiac muscle paralysis or neuromuscular paralysis. Toxins A and B cause rounding and eventual detachment of attached tissue culture cells [15].

This direct cytotoxicity is observed in many cell lines including intestinal epithelial cells, rat basophilic leukemia cells, ovarian and pancreatic cells and smooth muscle cells. Toxins A and B are stimulatory for mouse and human lymphocytes and macrophages *in vitro*, leading to cellular proliferation and release of cytokines [16]. Cell rounding is the basis of the tissue culture assay used in clinical laboratories to detect *C. difficile* toxins in stool samples. In human colonic tissues obtained from operative specimens both toxins cause diminished short circuit current, increased ionic permeability and surface colonocyte necrosis and apoptosis, with toxin B approximately 10 times more potent than toxin A [22]. Experiments using rodent ileal or colonic loops allow short-term (1 to 6 hr) study of toxin effects in the intact animal. In rabbit intestine toxin A, but not toxin B, is enterotoxic, eliciting massive fluid secretion, tissue necrosis and intense neutrophil accumulation in the lamina propria [23]. These inflammatory effects appear to result from activation of mast cells and sensory afferent neurons in the lamina propria [2, 19, 20]. Pharmacologic inhibition of mast cell degranulation and of substance P prior to or at the same time as toxin exposure block the inflammatory response and reduce intestinal secretion.

It is likely that the action of the toxins on the intact intestine involves two pathways as shown in Table 3. Direct effects of the toxins occur in enterocytes by binding of the toxin to receptors on the luminal border, internalization of toxin, inactivation of *rho* A and subsequent disaggregation of actin-containing filaments. Because little if any toxin can reach the lamina propria early in the course of enterotoxic action, it is likely that enterocyte products trigger the inflammatory reaction by initiating mast cell degranulation and release of substance P.

Table 3. Overview of enterotoxicity by *C. difficile* toxins

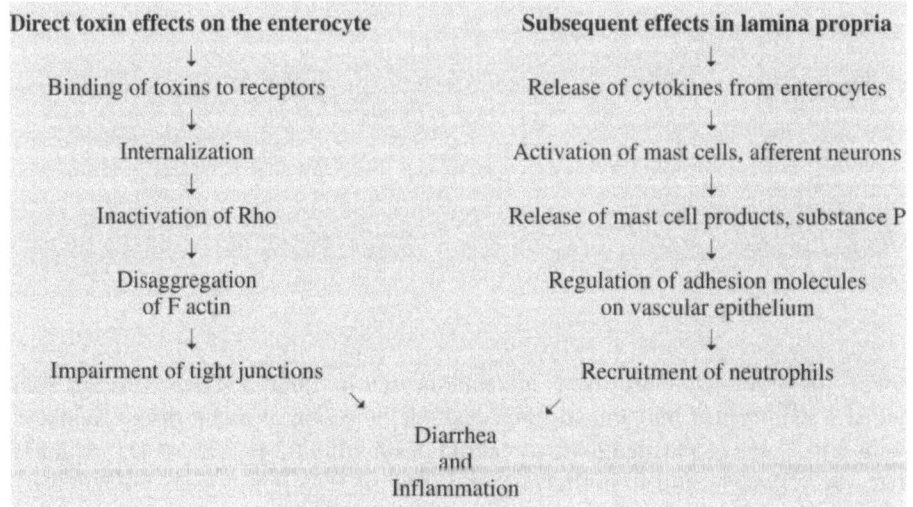

Direct toxin effects on the enterocyte	**Subsequent effects in lamina propria**
↓	↓
Binding of toxins to receptors	Release of cytokines from enterocytes
↓	↓
Internalization	Activation of mast cells, afferent neurons
↓	↓
Inactivation of Rho	Release of mast cell products, substance P
↓	↓
Disaggregation of F actin	Regulation of adhesion molecules on vascular epithelium
↓	↓
Impairment of tight junctions	Recruitment of neutrophils

Diarrhea
and
Inflammation

Toxin receptors

The biologic effects of all bacterial toxins require toxin binding to specific membrane receptors. Specific receptors for toxin A can be demonstrated on rabbit, hamster and human enterocyte membranes [3, 18]. The toxin A brush border receptor in rabbits and hamsters is a glycoprotein containing the trisaccharide Gal (alpha 1-3) Gal (beta 1-4) GlcNAc in the toxin binding domain. In contrast, the receptor on rabbit erythrocytes is a glycolipid containing the same trisaccharide. Mouse teratocarcinoma cells containing high levels of toxin-binding trisaccharide sequence are 100-times more sensitive to toxin A than cells which produce only small amounts of this trisaccharide on their surface. Human colonocytes possess specific receptors for toxin A but their binding activity is lower than observed with rabbit and rat enterocyte receptors. Recent results in our laboratory suggest that the brush border disaccharidase enzyme complex sucrase-isomaltase is a specific binding protein for toxin A. This enzyme is a glycoprotein containing terminal alpha-linked galactose residues that are required for toxin A binding. Clearly, sucrase is not present on colonocytes or other non-intestinal cells (e.g. fibroblasts, neutrophils) known to be sensitive to toxin A. Hence, other receptors bearing terminal alpha galactose residues must be present on these cells in order to explain toxin binding. Toxin A receptors are not expressed in newborn rabbit brush border [5] but slowly increase after weaning to adult levels. The absence of toxin A receptors in newborn rabbits might explain the relative resistance of newborns to toxin A. If similar developmental regulation of the toxin receptors occurs in humans, this might also explain the apparent resistance of human infants to *C. difficile*. Pretreatment of mice with monoclonal antibodies directed against the binding portion of the toxin A molecule protects these animals from experimental pseudomembranous colitis [14]. Thus, interference with toxin binding prevents enterotoxicity, and provides a rationale for vaccine development.

Intracellular effects of toxins

The biologic effects of toxin A on target cells appear to involve, at least in part, receptor-mediated activation of intracellular signaling mechanisms. For example, the toxin A receptor on rabbit brush borders [18] and neutrophil membranes [13]

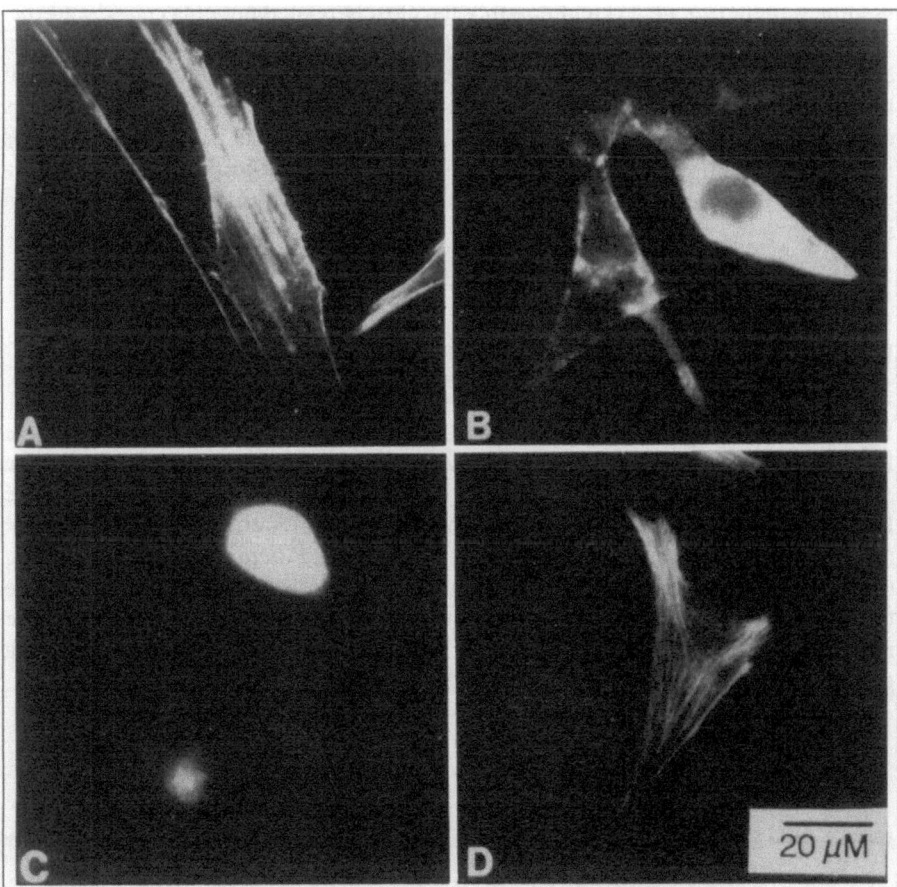

Fig. 1. NIH-3T3 fibroblasts were stained with rhodamine phalloidin to reveal filamentous actin, then exposed to toxin B (2 mg/ml). **Panel A** Control fibroblasts, no toxin. Note linear arrangement of actin filaments. **Panel B** After 15 minutes of toxin B exposure, actin filaments have disappeared. **Panel C** After 45 minutes exposure cell is rounded, and no filamentous actin can be seen. **Panel D** Cells are preincubated in calcium-free medium, then exposed to toxin B. Note no abnormality of actin filaments (reproduced from [17] by permission)

appears to be coupled to a pertussis toxin-sensitive heterotrimeric membrane G protein.

In addition, toxin A binding to its receptor in human neutrophils and rat pancreatic cells [8] caused a prompt increase in intracellular calcium and this increase was blocked by pertussis toxin. As shown in Fig. 1, exposure of 3T3 fibroblasts to toxin B causes progressive disaggregation of actin filaments (panels B and C) that can be blocked by preincubation of the cells in calcium-free medium to reduce intracellular calcium. These data indicate that intracellular calcium may be involved in the signal transduction of toxin A by a pathway which has not yet been defined.

After binding to its surface receptor toxins A and B enter the cell, possibly by endocytosis [7]. Depending on the toxin concentration, a lag period of 15 to 60 minutes occurs during which the toxin catalytic domain inactivates the GTP-binding protein *rho*, [4, 11] and possibly the closely related proteins *cdc* 42 and *rac*. These proteins, particularly *rho*, profoundly influence the dynamic equilibrium in cells between globular (G) and filamentous (F) actin [21]. Time course studies of human and rodent cells exposed to toxins A or B reveal loss of actin filaments as determined by staining with phallocidin [9, 17]. The identical action of both toxins in cells apparently relates to their ability to inactivate *rho* through a glucosyl-transferase reaction that adds a glucose residue to *rho*, rendering it unable to bind actin [12].

Future directions

Since the discovery of *C. difficile* as a major intestinal pathogen in 1977, we have learned a great deal about the structure and function of its toxins A and B. Yet the diseases related to these toxins are very common in hospitalized patients. In the United States alone it is estimated that more than 2 million cases occur each year, and most of these are in elderly, debilitated patients. Future efforts in controlling these diseases will require a better understanding of the host immune response to the toxins, as well as better insights into the complex changes of the colonic microflora following antibiotic therapy. Vaccination of high risk patients using genetically engineered toxins is one possible approach to controlling this disease.

References

1. Arnon SS, Mills DC, Day PA, Henrickson RV, Sullivan NM, Wilkins TD (1984) Rapid death of infant rhesus monkeys injected with *Clostridium difficile* toxins A and B: physiology and pathologic basis. J Pediatr 101: 34-40
2. Castagliuolo I, LaMont JT, Letourneau R, Kelly CP, O'Keane JC, Jaffer A, Theoharides TC, Pothoulakis C (1994) Neuronal involvement in the intestinal effects of *Clostridium difficile* toxin A and *Vibrio cholerae* enterotoxin in the rat. Gastroenterology 107: 657-665

3. Clark GF, Krivan NC, Wilkins TD, Smith BF (1987) Toxin A from *Clostridium difficile* binds to rabbit erythrocyte glycolipids with terminal Gal alpha 1-3Gal beta 1 4GlcNAc sequences. Arch Biochem Biophys 257: 217-229

4. Dillon S, Rubin E, Yakubovich M, Pothoulakis C, LaMont JT, Feig LA, Gilbert RJ (1995) The involvement of *ras*-related *rho* proteins in the mechanism of *Clostridium difficile* toxicity. Infect Immun 63: 1421-1426

5. Eglow R, Pothoulakis C, Itzkowitz S, Israel EJ, O'Keane C, Gong D, Gao N, Xu YL, Walker WA, LaMont JT (1992) Diminished *Clostridium difficile* toxin A sensitivity in newborn rabbit ileum is associated with decreased toxin A receptor. J Clin Invest 90: 822-829

6. vonEichel-Streiber C (1993) Molecular biology of the *Clostridium difficile* toxins. In: Sebald M (ed) Genetics and Molecular Biology of Anaerobic Bacteria. Springer Verlag, New York, pp 264-289

7. Fiorentini C, Thelestam M (1991) *Clostridium difficile* toxin A and its effects on cells. Toxicon 29: 543-567

8. Gilbert RJ, Pothoulakis C, LaMont JT, Yakubovich M (1995) *Clostridium difficile* toxin B activates calcium reflux which is required for actin disassembly during cytotoxicity. Am J Physiol (Gastrointest Liver Physiol) 31: G487-G494

9. Hecht G, Pothoulakis C, LaMont JT, Madara J (1988) *Clostridium difficile* toxin A perturbs cytoskeletal structure and tight junction permeability of cultured human intestinal epithelial monolayers. J Clin Invest 82: 1516-1524

10. Johnson JL, Phelps C, Barrioso L, Roberts M, Lyerly D, Wilkins TD (1990) Cloning and expression of the toxic B gene of *Clostridium difficile*. Current Microbiology 20: 397-401

11. Just I, Fritz G, Aktories K, Giry M, Popoff M, Bouquet P, Hegenbarth S, von-Eichel-Streiber C (1994) *Clostridium difficile* toxin B acts on the GTP-binding protein *rho*. J Biol Chem 269: 10706-10712

12. Just I, Selzer J, Wilm M, von Eichel-Streiber C, Mann M, Aktories K (1995) Glucosylation of *Rho* proteins by *Clostridium difficile* toxin B. Nature 375: 500-503

13. Kelly CP, Becker S, Linevsky JK, Joshi MA, O'Keane JC, Dickey BF, LaMont JT, Pothoulakis C (1994) Neutrophil recruitment in *Clostridium difficile* toxin A enteritis in the rabbit. J Clin Invest 93: 1257-1265

14. Lyerly D, Johnson J, Freg S, Wilkins TD (1990) Vaccination against lethal *Clostridium difficile* enterocolitis with a non-toxic recombinant peptide of toxin A. Current Microbiology 21: 29-32

15. Lyerly D, Krivan H, Wilkins TD (1991) *Clostridium difficile*: its diseases and toxins. Clin Microbiol Rev 88: 1-18

16. Miller PD, Pothoulakis C, Baeker TR, LaMont JT, Rothstein TL (1990) Macrophage-dependent stimulation of T cell depleted spleen cells by *Clostridium difficile* toxin A and calcium ionophore. Cell Immun 126: 155-163

17. Moore R, Pothoulakis C, LaMont JT, Carlson S, Madara JL (1990) *Clostridium difficile* toxin A increases intestinal permeability and induces Cl⁻ secretion. Am J Physiol (Gastrointest Liver Physiol) 22: G165-G172.

18. Pothoulakis C, LaMont JT, Eglow R, Gao N, Rubins JB, Theoharis TC, Dickey BF (1991) Characterization of rabbit ileal receptors for *Clostridium difficile* toxin A. Evidence for a receptor-coupled G-protein. J Clin Invest 88: 119-125

19. Pothoulakis C, Karmeli F, Kelly CP, Eliakim R, Joshi MA, O'Keane JC, LaMont JT, Rachmilewitz D (1993) Ketotifen inhibits *Clostridium difficile* toxin A-induced enteritis in rat ileum. Gastroenterology 105: 701-707

20. Pothoulakis C, Castagliuolo I, LaMont JT, Jaffer A, O'Keane JC, Snider RM, Leeman SE (1994) CP-96, 345, a substance P antagonist inhibits rat intestinal responses to *Clostridium difficile* toxin A, but not cholera toxin. Proc Natl Acad Sci 91: 947-951

21. Ridley A, Hall A (1992) The small GTP-binding protein *rho* regulates the assembly of focal adhesions and actin stress fibers in response to growth factors. Cell 70: 389-399

22. Riegler M, Sedivy R, Pothoulakis C, Hamilton G, Zacherl J, Bischof G, Cosentino E, Feil W, Schiessel R, LaMont JT, Wenz E (1995) *Clostridium difficile* toxin B is more potent than toxin A in damaging human colonic epithelium *in vitro*. J Clin Invest 95: 2004-2011

23. Triadafilopoulos G, Pothoulakis C, O'Brien M, LaMont JT (1987) Differential effects of *Clostridium difficile* toxins A and B on rabbit ileum. Gastroenterology 93: 273-279

24. Wren B (1991) A family of clostridial and streptococcal ligand-binding proteins with conserved C-terminal repeat sequences. Mol Microbiol 5: 797-803

Antibody response to *Clostridium difficile* toxins

Michel Delmée, Michel Warny

SUMMARY

Clostridium difficile A and B toxins induce the production of specific antibodies in both man and animals. In the hamster, passive oral immunization and vaccination against the two toxins prevents the development of a fatal colitis induced by *C. difficile*. The antibodies which are capable of neutralizing the enterotoxicity of toxin A recognise epitopes of the 38 repetitive sequences of the C-terminal end of the protein.

In man, serum IgG anti-toxin A and B, and serum and intestinal IgA anti-toxin A have been detected in two thirds of the normal population. The concentration of specific secretory IgA's was higher in the colonic mucosa than in the duodenum. Specific secretory IgA could inhibit *in vitro* the fixation of toxin A to the intestinal receptor of the rabbit. With symptomatic infection, a rise in IgG antitoxin A levels was seen in 11-68% of cases, and a rise in IgG antitoxin B in 44-71%. The serum and fecal IgA also rose to a significant extent. IgA was recognised to be responsible for the neutralising activity of the serum of convalescent patients.

Several authors have observed an association between the presence of relapsing colitis and a lack of increase in Ig anti-A antitoxin, though a causative relation remains unproven. In these patients, intravenous administration of immunoglobulin containing a high natural titer of IgG antitoxin A, was followed by a rapid and lasting clinical remission. The mechanism of action is unknown.

Improved knowledge of the roles of mucosal and systemic immunity against these toxins will allow to define the potential usefulness of a vaccine and the place of passive immunization in relapsing colitis.

Réponse anticorps aux toxines de *Clostridium difficile*

RÉSUMÉ

Les toxines A et B de *C. difficile* entraînent la production d'anticorps spécifiques, chez l'homme comme chez l'animal. Chez le hamster, l'immunisation orale passive et la vaccination contre les deux toxines préviennent le développement d'une colite létale induite par *C. difficile*. Les anticorps capables de neutraliser l'entérotoxicité de la toxine A reconnaissent certains épitopes, parmi les 38 séquences répétitives de l'extrémité carboxyl-terminale de la protéine.

Chez l'homme, les IgG sériques anti-toxines A et B et les IgA sériques et intestinales anti-toxines A ont été détectées dans deux tiers de la population normale. La concentration des IgA sécrétoires (IgAs) était plus élevée dans la muqueuse colique que dans le duodénum. Les IgAs pouvaient inhiber *in vitro* la fixation de la toxine A au récepteur intestinal du lapin.

Lors d'une infection symptomatique, une élévation des IgG anti-toxine A est notée dans 11 à 68 % des cas, et une élévation des IgG anti-toxine B dans 44 à 71 % des cas. Les IgA sériques et fécales s'élèvent également de façon significative. Les IgA sont considérées comme responsables de l'activité neutralisante du sérum des patients convalescents.

Plusieurs auteurs ont observé une association entre la présence de colites récidivantes et une absence d'augmentation des IgA anti-toxine A, bien qu'une relation de cause à effet demeure non établie. Chez ces patients, l'administration intraveineuse d'immunoglobulines contenant un titre naturellement élevé en IgG anti-toxine A, fut suivie d'une rémission clinique rapide et durable. Les mécanismes d'action sont inconnus.

L'amélioration des connaissances sur le rôle de l'immunité locale et systémique contre les toxines de *C. difficile* permettra de définir l'utilité potentielle d'un vaccin et la place de l'immunisation passive dans les colites récidivantes.

Antikörper-Reaktion auf *Clostridium difficile*-Toxine

ZUSAMMENFASSUNG

Die *Clostridium difficile*-Toxine A und B induzieren bei Menschen und Tieren die Bildung spezifischer Antikörper. Beim Hamster verhütet die passive orale Immunisierung und Impfung gegen die beiden Toxine die Entstehung einer durch eine Infektion mit *C. difficile* ausgelösten letalen Kolitis. Antikörper, die in der Lage sind, die Enterotoxizität von Toxin A zu neutralisieren, erkennen bestimmte Epitope der 38 repetitiven Sequenzen am Carboxylende des Proteins.

Beim Menschen wurden IgG-Antikörper gegen die Toxine A und B im Serum sowie IgA-Antikörper gegen Toxin A im Serum und im Darm bei zwei Dritteln der gesunden Bevölkerung nachgewiesen. Die Konzentration an spezifischem sekretorischen IgA war in der Colonmucosa höher als im Duodenum. Spezifisches sekretorisches IgA hemmte *in vitro* die Bindung von Toxin A an die Rezeptoren im Kaninchendarm. Bei symptomatischen Infektionen wurde in 11% - 68% der Fälle ein Anstieg der IgG-Antitoxin A-Spiegel sowie in 44% - 71% der Fälle ein Anstieg des IgG-Antitoxin B registriert. Die IgA-Spiegel in Serum und Faeces stiegen ebenfalls in beträchtlichem Maße an. IgA wird als Ursache für die neutralisierende Wirkung des Serums genesender Patienten angesehen.

Mehrere Autoren bemerkten einen Zusammenhang zwischen dem Vorliegen einer rezidivierenden Kolitis und einem mangelnden Anstieg des IgA-Antitoxins A; allerdings ist eine kausale Beziehung noch unbewiesen. Bei diesen Patienten folgte auf die intravenöse Gabe von Immunglobulin mit einem hohen natürlichen Titer von IgG-Antitoxin A eine rasche und dauerhafte klinische Remission. Der Wirkmechanismus ist unbekannt.

Eine bessere Kenntnis der Rolle der mukosalen und systemischen Immunität gegen diese Toxine wird in die Lage versetzen, den potentiellen Wert entsprechender Impfungen und die Rolle der passiven Immunisierung gegen rezidivierende Colitiden zu würdigen.

Introduction

The pathogenicity of *Clostridium difficile* is linked to the production of two protein exotoxins, known as A and B. In vivo, these differ in their effect on the intestinal mucosa. In rodents, oral administration of toxin A produces a type of colitis characterized by a neutrophil infiltration of the mucosa, the production of an inflammatory exudate rich in serum immunoglobulins and necrosis of the intestinal epithelium. In contrast, toxin B is not enterotoxic in animals [11, 20].

Toxins A and B induce the production of specific antibodies, in animals as well as in man. We will try to summarise the data on the characteristics and role of these antibodies.

Immunization against toxins A and B in animals

Active immunization against the toxins of *C. difficile* has been studied mainly in the hamster and the rabbit. Libby et al were the first to observe that parenteral immunization against toxins A or B induced the appearance of specific antibodies in the serum [17]. In contrast to vaccination against one of these toxins, immunization against both prevented the appearance of a lethal colitis [17]. Nonetheless, Kim et al observed that 100% of hamsters immunized by the subcutaneous route against toxin A or both toxins were protected against *C. difficile* colitis induced by clindamycin, whereas no animals immunized against toxin B only, survived [14]. Furthermore, the maternal milk of adults immunized against toxin A or both toxins protected 48% and 28% of young hamsters, respectively [14]. Later, these authors were able to show that the antitoxin activity of the milk and the intestinal mucosa was dependent on IgG [15].

In the rabbit, Ketley et al observed that systemic immunization against toxin A reduced the intensity of the mucosal lesions [12]. This protection was less following the simultaneous administration of both toxins. Wilkin's group observed that a monoclonal antitoxin A (PCG-4) which was unable of neutralizing the cytotoxic action of toxin A on CHO and T84 cells, could nonetheless neutralize its enterotoxicity in the rabbit [18, 19].

The monoclonal PCG-4 recognises two regions of toxin A located in the C-terminal end of the protein [7]. This non-toxic portion comprises 38 segments with a high nucleotide sequence similarity. It is responsible for red cell agglutination in the rabbit and probably links the toxin to its intestinal receptor. Immunization against this region protects the animals against a lethal colitis [21]. Recently, sucrase-isomaltase has been identified as the receptor for toxin A in the small intestine of the rabbit [JT LaMont, In "Updates on *C. difficile*", Symposium of 5 May 1995 Paris]. This enzyme is not expressed by human colonic epithelium.

Passive immunization by the intravenous route has been evaluated in the mouse by Corthier et al, using different monoclonal antibodies against toxin A, including PCG-4 [4, 5]. All of these recognise the repeated units at the carboxyl end of the protein. Intravenous administration of these monoclonal antibodies protected 100% of individuals against a lethal colitis induced by the VP1 10463 strain. The treated mice had a level of intestinal A toxin ten times less than the controls, whereas the level of toxin B was the same in both groups.

Antibodies against toxins A and B in the normal population

Serum immunoglobulins

Several authors have reported that some two thirds of adults have antitoxin A IgG (range 57%-72%) and antitoxin B IgG (66%) [10, 26, 28]. In contrast, only 19% of individuals of less than 2 years of age are positive for antitoxin A IgG, showing that immunization occurs naturally during the first few years of life [26]. This significant difference has been observed by others [10, 28]. Very high ELISA titers correlated well with neutralizing activity *in vitro* [26]. A similar incidence was found in respect of serum antitoxin A IgA (51%-60%) [10, 28].

Secretory IgA (S-IgA) antitoxin A

Wada et al have observed that the aqueous phase of the colostrum of 28% of Japanese women (n = 60) could neutralise *in vitro* the cytotoxic effect of a preparation of unpurified toxin. This neutralizing activity was inhibited by IgA anti-A antibodies [27]. Kim et al have also shown a neutralizing activity in the colostrum of 56% of American women (n = 55) with respect to toxins A and B. The titer of this activity correlated well with the protective effect of the administration of colostrum to mice [13].

S-IgA anti-toxin A was detected in 70% and 40% of the stools of adults, and of children below the age of two, respectively [28]. At the same time, S-IgA specific for toxin A has been found in the colonic aspirates of healthy adults [10]. The antibody level was significantly lower in the duodenum. Samples containing high levels of S-IgA inhibited the fixation of toxin A to the intestinal receptor in the rabbit [10]. These findings may explain the presence of toxin B in the stools of asymptomatic carriers.

Antibodies against toxin A and toxin B in colitis

Serum immunoglobulins

Individuals vary widely in the intensity of their humoral response to toxins A and B, which is influenced by any coexisting pathology and also by treatment with antibiotics. A serum response of IgG antitoxin A was seen in 11 to 68%, and a rise in antitoxin B IgG in 44 to 71% of cases [1, 2, 26, 28].

It was significantly less frequent in patients receiving chemotherapy [28]. The neutralizing activity of serum IgG *in vitro* also rose significantly [8, 26]. Recently, Johnson et al have analysed the neutralizing activity of antitoxin A in the serum of one third of convalescent patients (n = 18) [9]. In spite of the presence of high titers of antitoxin A IgG, the neutralizing activity on OTF9-63 cells and on the rabbit ileal loop [3] was due to the IgA fraction of the serum immunoglobulins. A strong response to IgA antitoxin A was sometimes seen in the absence of a rise in specific IgG [8, 28].

IgG directed against the surface antigens of *C. difficile* were found in 50% of a series of 15 patients presenting with antibiotic associated diarrhea [24].

Fecal IgA antitoxin A

Symptomatic infection induces a significant rise in total IgA and antitoxin A IgA levels [8, 28]. Estimation of their levels by immunoassay is very inaccurate in the absence of a separation of the different molecular forms of IgA, the proportions of which vary during the course of intestinal inflammation [6, 23]. Concentrations of total IgG in the region of 1 mg/ml have been measured in the stools of a small proportion of patients. As Corthier et al have shown in the mouse [5], a variable fraction of the immunoglobulins found in the stools of patients with diarrhea, originates from exudation of plasma.

Relation between antibody levels and relapsing colitis

Infection with *C. difficile* is accompanied by clinical manifestations of varying severity, ranging from asymptomatic carrier status to pseudomembranous colitis. The causes of this variability are poorly understood, but host factors play an important part [22]. Recurrence is seen in 10 to 20% of cases, of unknown origin. Repeatedly recurrent and fulminant forms of the disease are difficult to treat [11].

Several authors have therefore evaluated the humoral response to the toxins in patients suffering from recurrent diarrhea. In six children with this complaint, Leung et al observed a level of antitoxin A IgG significantly lower than that of a normal population adjusted for age [16]. The levels of specific IgA were however similar in the two groups. In four adults suffering from recurrent colitis, we observed levels of serum IgG and fecal antitoxin A IgA that were significantly lower than in those who presented with a single episode (Fig. 1) [28]. In contrast, Johnson et al found serum levels of IgA and IgG in three adults with recurrences to be the same as in those with no recurrence [8]. In seven patients with recurrent diarrhea, Bacon et al observed only one case who tested positive for antitoxin A and antitoxin B, and one other positive for antitoxin B alone [2]. Finally, in eight patients with recurrence, we saw no response to toxin A, but four of these had very high levels of IgG antitoxin B [unpublished data].

In all, these findings suggest an association between the presence of recurrent colitis and a lack of humoral response to toxin A. A causative relation has none-theless not as yet been proven.

Passive immunization in relapsing colitis

Leung et al administered human immunoglobulins from pooled donors to five children presenting with relapsing colitis associated with very low levels of antitoxin A IgG [16]. This therapeutic trial was based on the known protective effect of antitoxin antibody in animals, and also, on the presence of specific IgG in two thirds of subjects. These authors observed a significant rise in the titer of serum antitoxin A IgG together with a remission in clinical symptoms. This was accompanied by a disappearance of toxin B from the stools in four of the patients. We have applied this treatment to one patient who had experienced five recurrent

Antitoxin A antibody levels

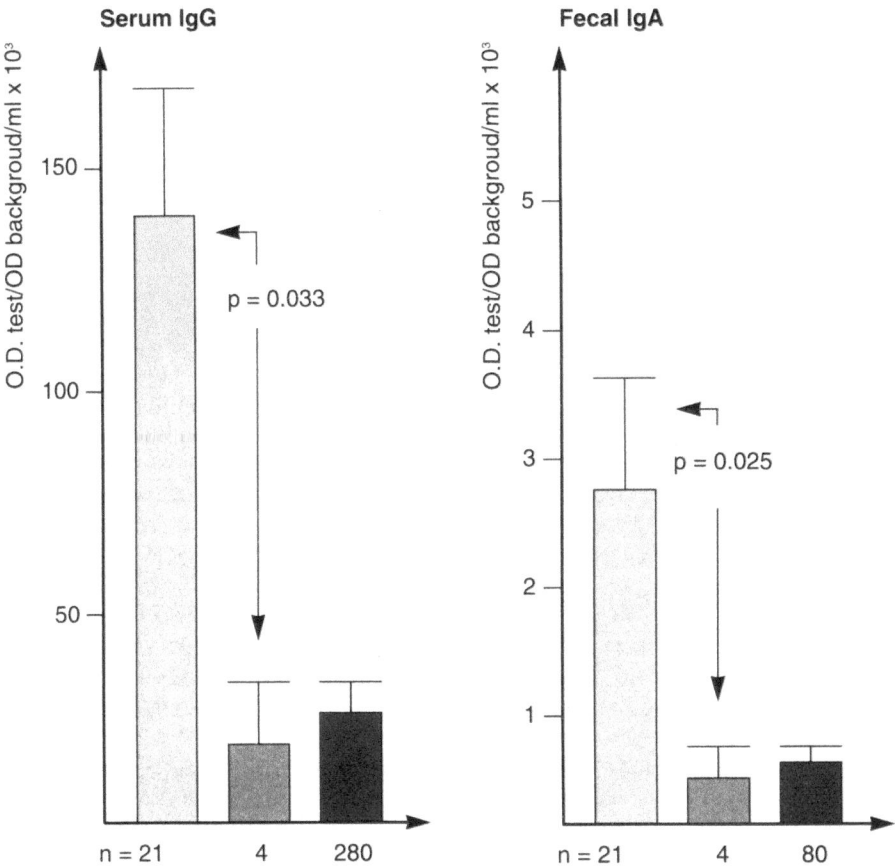

Fig. 1. Antitoxin A serum IgG and fecal IgA in controls (n = 280 and 80, respectively) and in *C. difficile* diarrhea. Antibody levels were significantly higher in patients who presented a single episode (n = 21) than in those with relapsing diarrhea (n = 4). Results were expressed as the mean ± standard error. Differences in antibody levels were assessed using the Mann-Whitney U test. Warny M et al [28]

episodes, with the same result (Fig. 2) [29]. Recently, four additional patients with relapsing disease and no antibody response to toxin A were cured after having received this treatment [unpublished data]. Oral administration of IgA was reported in a single patient [25].

Based on a somewhat limited experience, the intravenous administration of immunoglobulins seems to be effective in patients with recurrent colitis who have an absent response to serum antitoxin A. The mode of action is unknown, and the role of antitoxin IgG remains to be demonstrated. A controlled trial using two pools

Intravenous gammaglobulin

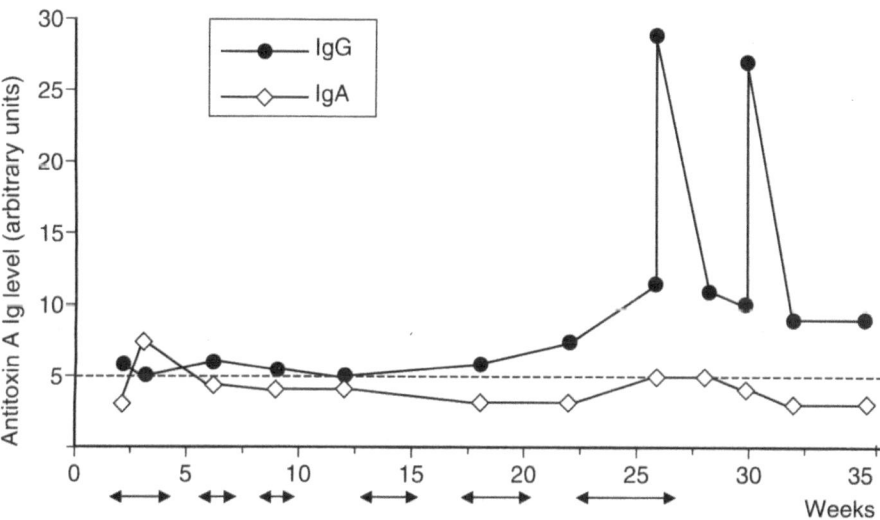

Fig. 2. Time course of serum antitoxin A IgG and IgA levels in a patient who suffered five episodes of pseudomembranous colitis. Double arrows show when the patient was hospitalized for colitis, and the dashed line indicated the cutoff of ELISA. During the fifth episode, two doses of intravenous gammaglobulin were administered (400 mg/kg), in addition to vancomycin therapy. The patient became asymptomatic after the first dose and remained negative for serum antitoxin A antibodies (follow-up: 3 years). Warny et al [29]

of immunoglobulin, one from seropositive and the other from seronegative donors, for antitoxin A IgG, could help to clarify their role.

Conclusions and perspectives

The protective role of the antibody response to *C. difficile* toxins, in particular to toxin A, has been demonstrated in animal models. In humans, this hypothesis, although suggested by several data, should be investigated in a large prospective study of the incidence of the *C. difficile* infection in relation to antibody levels.

Although these have only been attempted on a small scale, trials of passive immunization in severe or recurrent cases seem to be an approach that deserves more detailed appraisal. A better understanding of the role of systemic and mucosal antibodies will allow us to discern more precisely the place of passive immunization in the severe and recurrent forms of the disease, and possibly to design a future vaccine.

References

1. Aronsson B, Granström M, Möllby R, Nord, CE (1985) Serum antibody response to *Clostridium difficile* toxins in patients with *Clostridium difficile* diarrhoea. Infection 13: 97-101

2. Bacon AE III, Fekety R (1994) Immunoglobulin G directed against toxins A and B of *Clostridium difficile* in the general population and patients with antibiotic-associated diarrhea. Diagn Microbiol Infect Dis 18: 205-209

3. Bartlett JG, Tedesco FJ, Shull S, Lowe B, Chang TW (1980) Symptomatic relapse after oral vancomycin therapy of antibiotic-associated pseudomembranous colitis. Gastroenterology 78: 431-434

4. Corthier G., Muller MC, Wilkins TD, Lyerly D, L'Haridon R (1991) Protection against experimental pseudomembranous colitis in gnotobiotic mice by use of monoclonal antibodies against *Clostridium difficile* toxin A. Infect Immun 59: 1192-1195

5. Corthier G, Muller MC, Elmer GW, Lucas F, Dubos-Ramaré F (1989) Inter-relationships between digestive proteolytic activities and production and quantitation of toxins in pseudomembranous colitis induced by *Clostridium difficile* in gnotobiotic mice. Infect Immun 57: 3922-3927

6. Delacroix D, Dehennin JP, Vaerman JP (1982) Influence of molecular size of IgA on its immunoassay by various techniques. II. Solid phase radioimmunoassays. J Immunol Methods 48: 327-337

7. Frey JM, Wilkins TD (1992) Localization of two epitopes recognized by monoclonal antibody PCG-4 on *Clostridium difficile* toxin A. Infect Immun 60: 2488-2492

8. Johnson S, Gerding DN, Janoff EN (1992) Systemic and mucosal antibody responses to toxin A in patients infected with *Clostridium difficile*. J Infect Dis 166: 1287-1294

9. Johnson S, Sypura WD, Gerding DN, Ewing SL, Janoff EN (1995) Selective neutralization of a bacterial enterotoxin by serum immunoglobulin A in response to mucosal disease. Infect Immun 63: 3166-3173

10. Kelly CP, Pothoulakis C, Orellana J, LaMont JT (1992) Human colonic aspirates containing immunoglobulin A antibody to *Clostridium difficile* toxin A inhibit toxin A-receptor binding. Gastroenterology 102: 35-40

11. Kelly CP, Pothoulakis C, LaMont JT (1994) *Clostridium difficile* colitis. (Review article). N Engl J Med 330: 257-262

12. Ketley JM, Mitchell TJ, Candy DCA, Burdon DW, Stephen J (1987) The effect of *Clostridium difficile* crude toxins and toxin A on ileal and colonic loops in immune and non-immune. J Med Microbiol 24: 41-52

13. Kim K, Pickering LK, DuPont HL, Sullivan N, Wilkins TD (1984) *In vitro* and *in vivo* neutralizing activity of human colostrum and milk against purified toxins A and B of *Clostridium difficile*. J Infect Dis 150: 57-62

14. Kim PH, Iaconis JP, Rolfe RD (1987) Immunization of adult hamsters against *Clostridium difficile*-associated ileocecitis and transfer of protection to infant hamsters. Infect Immun 55: 2984-2992

15. Kim PH, Rolfe RD (1989) Characterisation of protective antibodies in hamsters immunised against *Clostridium difficile* toxins A and B. Microb Ecol Health Dis 2: 47-59

16. Leung DYM, Kelly CP, Boguniewicz M, Pothoulakis C, LaMont JT, Flores A (1991) Treatment with intravenously administered gamma globulin of chronic relapsing colitis induced by *Clostridium difficile* toxin. J Pediatr 118: 633-637

17. Libby JM, Jortner BS, Wilkins TD (1982) Effect of the two toxins of *Clostridium difficile* in antibiotic-associated cecitis in hamsters. Infect Immun 36: 822-829

18. Lima AAM, Lyerly DM, Wilkins TD, Innes DJ, Guerrant RL (1988) Effects of *Clostridium difficile* toxins A and B in rabbit small and large intestine *in vivo* and on cultured cells *in vitro*. Infect Immun 56: 582-588

19. Lyerly DM, Phelps CJ, Toth J, WilkinsTD (1985) Monoclonal and specific polyclonal antibodies for immunoassay of *Clostridium difficile* toxin A. J Clin Microbiol 21: 12-14

20. Lyerly DM, Krivan HC, Wilkins TD (1988) *Clostridium difficile*: its disease and toxins. Clin Microbiol Rev 1: 1-18

21. Lyerly DM, Johnson JL, Frey SM, Wilkins TD (1990) Vaccination against lethal *Clostridium difficile* enterocolitis with a non toxigenic recombinant peptide of toxin A. Curr Microbiol 21: 29-32

22. McFarland LV, Elmer GW, Stamm WE, Mulligan ME (1991) Correlation of immunoblot type, enterotoxin production and cytotoxin productionwith clinical manifestations of *Clostridium difficile* infection in a cohort of hospitalized patients. Infect Immun 59: 2456-2462

23. Meillet D, Raichwarg D, Tallet F, Savel J, Yonger J, Gobert J (1987) Measurement of total, monomeric, and polymeric IgA in human feces by electroimmunodiffusion. Clin Exp Immunol 69: 142-147

24. Pantosti A, Cerquetti M, Viti F, Ortisi G, Mastrantonio P (1989) Immunoblot analysis of serum immunoglobulin G response to surface proteins of *Clostridium difficile* in patients with antibiotic-associated diarrhea. J Clin Microbiol 27: 2594-2597

25. Tjellström B, Stenhammar L, Eriksson S, Magnusson KE (1993) Oral immunoglobulin A supplement in treatment of *Clostridium difficile* enteritis. Lancet 341: 701-702

26. Viscidi R, Laughon BE, Yolken R, Bo-Linn P, Moench T, Ryder RW, Bartlett JG (1983) Serum antibody response to toxins A and B of *Clostridium difficile*. J Infect Dis 148: 93-101

27. Wada N, Nishida N, Iwaki S, Ohi H, Miyawaki T, Taniguchi N, Migita S (1980) Neutralizing activity against *Clostridium difficile* toxin in the supernatants of cultured colostral cells. Infect Immun 29: 545-550

28. Warny M, Vaerman JP, Avesani V, Delmée M (1994) Human antibody response to *Clostridium difficile* toxin A in relation to clinical course of infection. Infect Immun 62: 384-389

29. Warny M, Denie C, Delmée M, Lefebvre C (1995) Gamma-globulin administration in relapsing *Clostridium difficile*-induced pseudomembranous colitis with a defective antibody response to toxin A. Acta Clin Belg 50: 36-39

Laboratory methods for detecting the toxins of *Clostridium difficile*

Frédéric Barbut, Jean-Claude Petit

SUMMARY

Clostridium difficile is the principal gut pathogen responsible for nosocomial diarrhea in the adult. This anaerobic organism is isolated in 10% of stool cultures received in hospital microbiology laboratories. Many methods and strategies have been devised for the diagnosis of *C. difficile* related infection. Among these, the detection of the toxins (which are directly implicated in the pathophysiology of the digestive problems caused by *C. difficile*) have been considered the best means of diagnosis. Current practice in this area has developed considerably over the last ten years.

The detection of toxin B by the stool cytotoxicity assay (SCA) is still considered today to be the standard method. Whatever cell line is used (CHO, HeP-2, MRC5, Vero), it provides good sensitivity (detecting levels of toxin B of the order of a picogram) and excellent specificity when the cytopathic effect is neutralized by a specific antiserum. However this technique suffers from several disadvantages including time (it takes 24-48 hours), lack of standardization, heavy demands on staff, and difficulty in obtaining the antiserum.

Identification of the toxins has recently been improved by the introduction of commercial immuno-enzymatic (ELISA) tests, which detect either toxin A alone or both A and B simultaneously. The specificity of these tests is excellent (> 97% as a rule) but their sensitivity varies from 52% to 99%. Results are available in less than three hours which may allow the time to begin necessary hygienic precautions, so as to limit the spread of infection of these organisms. These tests, in spite of their high unit costs, represent an excellent alternative for those bacteriology laboratories which lack the facilities required for cell culture.

More recently, cloning and gene sequencing of toxins A and B have opened the way for molecular biology techniques. Membrane hybridization of the stools with the aid of specific probes for toxin B, have shown a 96% correlation with the SCA. Polymerase chain reaction (PCR) has also been applied to the detection of the genes of toxins A and B directly from the stools. The sensitivity of this method is higher than the SCA, but it is more difficult to perform and the many polymerase inhibitors present in the stools may interfere with the PCR. To avoid these snags, a magnetic Immuno PCR assay (MIPA) has recently been introduced. It consists in capturing *C. difficile* with magnetic spheres coated with antibody, then separating them from the stools so as to amplify the toxin B gene by PCR. Methods of detection of *C. difficile* toxins A and B by molecular biology are at a very early stage, and their role in relation to the established tests remains to be defined.

Méthodes de détection des toxines de *Clostridium difficile* au laboratoire

RÉSUMÉ

C. difficile est le principal entéropathogène bactérien responsable de diarrhées nosocomiales chez l'adulte. Ce microorganisme anaérobie est isolé dans 10 % des coprocultures reçues dans des laboratoires hospitaliers de microbiologie. De nombreuses méthodes et techniques ont été mises au point pour le diagnostic des infections dues à *C. difficile*. Parmi elles, la détection des toxines (qui sont impliquées directement dans la physiopathologie des troubles digestifs causés par *C. difficile*) est considérée comme le meilleur moyen de diagnostic. Dans ce domaine, la pratique de routine s'est considérablement développée depuis ces dix dernières années.

La détection de la toxine B par la recherche d'un effet cytopathogène sur un filtrat de selles est considérée aujourd'hui encore comme la méthode de référence (test de cytotoxicité). Quelle que soit la lignée cellulaire utilisée (CHO, HeP-2, MRC5, Vero), elle possède une bonne sensibilité (seuil de détection de la toxine B de l'ordre du picogramme), et une excellente spécificité quand l'effet cytopathogène est neutralisé par un antisérum spécifique. Cependant cette technique souffre de plusieurs inconvénients, notamment le délai de réalisation (24 à 48 heures), le manque de standardisation, la nécessité d'une lourde infrastructure et la difficulté pour obtenir les antisérums.

L'identification des toxines a récemment été améliorée par la commercialisation de tests immunoenzymatiques (de type ELISA), qui détectent soit la toxine A isolée, soit les deux toxines A et B simultanément, à l'aide d'anticorps monoclonaux. La spécificité de ces tests est excellente (> 97 % en règle générale), mais leur sensibilité varie de 52 à 99 %. Les résultats sont disponibles en moins de trois heures, ce qui permet un traitement spécifique précoce et l'instauration rapide de mesures d'hygiène indispensables pour limiter la transmission nosocomiale de cette bactérie. Ces tests, en dépit de leur coût unitaire élevé, représentent une excellente alternative pour les laboratoires de bactériologie dépourvus de l'infrastructure nécessaire à la culture cellulaire.

Plus récemment, le clonage et le séquençage des gènes des toxines A et B ont ouvert la voie des techniques de biologie moléculaire. L'hybridation à l'aide de sondes spécifiques de la toxine B, d'une suspension de selles déposée sur membrane a montré une corrélation de 96 % avec le test de cytotoxicité des selles. La réaction de polymérisation en chaîne (PCR) a aussi été appliquée à la détection des gènes des toxines A et B, directement à partir des selles. La sensibilité de cette méthode est supérieure à celle du test de cytotoxicité des selles, mais elle est plus délicate à réaliser, et les nombreux inhibiteurs de la polymérase présents dans les selles peuvent interférer avec la PCR. Pour éviter ces écueils, une technique de PCR sensibilisée par procédé immuno-magnétique, a récemment été élaborée. Elle consiste en la capture de *C. difficile* par des sphères magnétiques recouvertes d'anticorps, qui sont ensuite séparées des selles pour permettre d'amplifier le gène de la toxine B par PCR. Les méthodes de détection des toxines A et B de *C. difficile* par biologie moléculaire en sont à un stade très précoce de leur développement, et leur rôle par rapport aux tests usuels de diagnostic devra, à l'avenir, être défini.

Labormethoden zum Nachweis der *Clostridium difficile*-Toxine

ZUSAMMENFASSUNG

Clostridium difficile ist das wichtigste intestinale Pathogen und ist verantwortlich für nosokomiale Diarrhöen beim Erwachsenen. Der anaerobe Mikroorganismus läßt sich aus 10% der in mikrobiologischen Labors von Krankenhäusern untersuchten Stuhlproben kultivieren. Für die Diagnostik *C. difficile*-assoziierter Infektionen wurden zahlreiche Methoden und Strategien entwickelt. Von diesen gilt der Nachweis der *C. difficile*-Toxine (die direkt an der Pathophysiologie von Verdauungsstörungen im Zusammenhang mit *C. difficile* beteiligt sind) als zuverlässigstes diagnostisches Werkzeug. Die gängige Praxis auf diesem Gebiet hat sich in den letzten zehn Jahren erheblich weiterentwickelt.

Der Nachweis von Toxin B mittels zytopathischer Effekte (CPE) von Stuhlfiltraten gilt heute noch als Standardmethode. Bei allen verwendeten Zellinien (CHO, HeP2, MRC5, Vero) bietet der Test eine gute Sensitivität (die Nachweisgrenze von Toxin B liegt im Picogrammbereich) und eine hervorragende Spezifität durch Neutralisation der zytopathischen Effekte durch ein spezifisches Antiserum. Allerdings weist das Verfahren mehrere Nachteile auf, darunter den Zeitaufwand (Dauer 24-48 Stunden), die fehlende Standardisierung, die starke Belastung des Personals und Probleme bei der Beschaffung des Antiserums.

Der Nachweis der Toxine ist erleichtert, seit neuerdings immun-enzymatische (ELISA) Tests, die entweder nur Toxin A oder gleichzeitig Toxin A und B nachweisen, im Handel sind. Die Spezifität dieser Tests ist ausgezeichnet (in der Regel > 97%), allerdings schwankt ihre Sensitivität zwischen 52% und 99%. Die Ergebnisse liegen in weniger als drei Stunden vor, wodurch Zeit für hygienische Vorkehrungen gewonnen wird, um die Ausbreitung von Infektionen mit diesen Keimen zu verhüten. Die Tests stellen trotz ihrer hohen Kosten eine hervorragende Alternative für solche bakteriologische Labors dar, denen die für Zellkulturen erforderlichen Einrichtungen fehlen.

Clostridium difficile is considered to be the principal gut pathogen responsible for nosocomial diarrhea in the adult [13, 17, 26]. The clinical spectrum of gut infections associated with this spore forming anaerobic organism varies from moderate diarrhea to fulminating pseudomembranous colitis [21]. The pathogenicity of *C. difficile* is essentially due to the simultaneous production of two toxic proteins [3, 39], known as toxin A (enterotoxin) and toxin B (cytotoxin). Following breakdown of the natural defences of the alimentary tract, favoured by antibiotics, toxin A becomes fixed to a glycoprotein receptor on the brush border and induces an intense inflammatory response, with a hemorrhagic exudate. In synergy with toxin A, toxin B exerts a powerful cytotoxic action which ends in total destruction of the cytoskeleton.

Over the last 15 years many strategies have been proposed to establish the diagnosis of *C. difficile* infection [6, 7]. These include endoscopy to reveal pseudomembranes, detection of metabolites such as isocaproic acid or specific antigens of *C. difficile* in the stools (agglutination tests) [20, 23, 30], bacterial isolation of *C. difficile* on selective culture media or the direct detection of the toxins in the stools.

Demonstration of toxin in the stools has long been favoured by bacteriologists as it permits the direct detection of the virulent factors of *C. difficile*. Knowing that all toxinogenic strains of the organism secrete both toxins A and B, their detection in the stool is considered to be an accurate marker of the presence of a pathogenic strain. This approach has been stimulated by the recent introduction of commercial kits for the immuno-enzymatic detection of toxins A and B.

History

Detection of the toxins of *C. difficile* in the stools has been developed by the use of four successive techniques (Table 1), namely stool cytotoxicity assay (SCA), counterimmunoelectrophoresis (CIE), immuno-enzymatic (ELISA) tests, and techniques using molecular biology (hybridization or polymerase chain reaction (PCR). CIE was described by several teams in the early 1980's, but has now become obsolete because of its lack of sensitivity and specificity [31, 37, 44].

SCA has been in use since 1978 and immuno-enzymatic tests since 1990, and both tests are widely used in microbiology laboratories. A study carried out in 1995 which included 82 hospital laboratories from every part of France showed that 57% of them used toxin detection for the diagnosis of *Clostridium difficile* infection. More than three quarters used ELISA and 25% cell culture [personal communication].

Molecular biology applied to toxin detection has been developed since 1991, but is still a research tool.

Whatever method is used leads in most cases to the rapid identification of toxin in the stools. Nonetheless, some laboratories prefer to culture the organism as a first stage, and to determine its toxin producing properties at a second stage. Although this method gives excellent sensitivity as regards the diagnosis of *C. difficile*

Table 1. Methods for detection of toxins of *C. difficile*

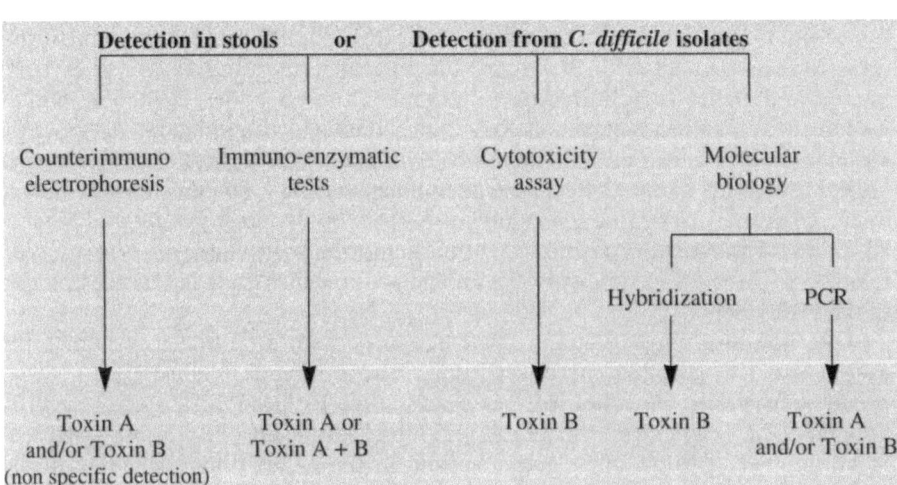

infection, it is somewhat tedious (minimum 5 days) and unsuitable to the emergency situation.

The stool cytotoxicity assay (SCA) (Fig. 1)

The principle of the SCA initially described by Chang et al [8] in 1978, relies on the powerful cytotoxic action of toxin B in cell culture. In fact, toxins A and B are both cytotoxic, but the activity of toxin B is estimated to be 10^3 to 10^4 times greater than that of toxin A. In practice, the test consists in inoculating a sterile stool filtrate into a cell line and after 24 to 72 hours incubation at 37° C in a CO_2 enriched atmosphere measuring the cytopathic effect which is neutralized by a specific antiserum against *C. difficile* or *C. sordellii*. Many cell lines can be used, including human or animal fibroblasts (MRC-5, WI-38, L929), renal cells (AGMK, Vero, BHK-21) or CHO-K, McCoy, HeLa, or HeP-2 cells [25, 29, 35].

They differ from one another in the ease of identification of cytopathic effect (MRC-5 cells undergo a characteristic "globulisation") and in the delay of appearance. The detection threshold of these methods varies from one line to another but is in the order of a picogram of toxin B [25, 29]. The absence of correlation between the amounts of toxin B in the stools and the severity of the symptoms means that the level of toxin should not be quantified and the result is expressed simply as « positive » or « negative ».

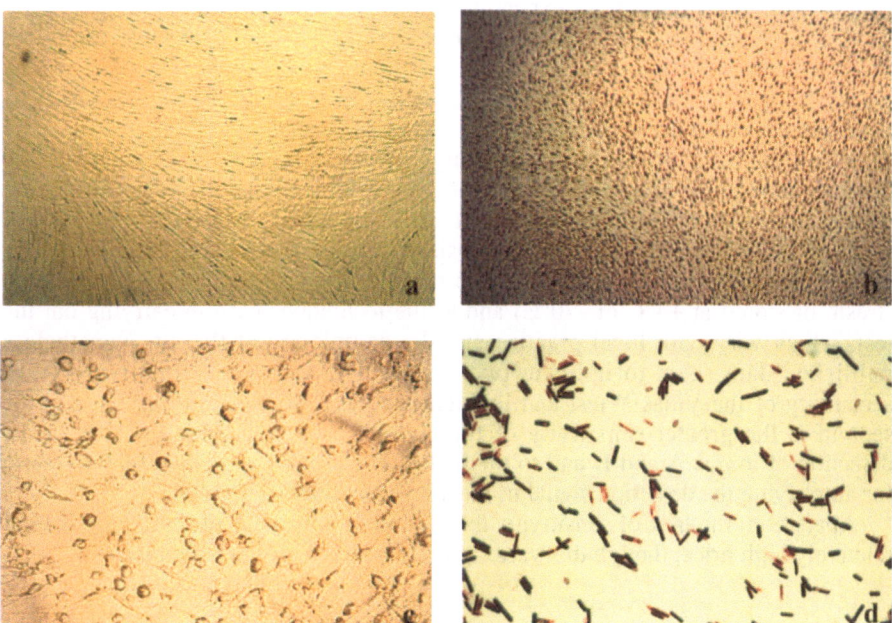

Fig. 1. a Culture of MRC5 (x 25). **b** Cytopathic effect of *C. difficile* toxin B on MRC5 cells (x 25). **c** Cytopathic effect of *C. difficile* toxin B on MRC5 cells (x 200). **d** Gram stain preparation of *C. difficile* (x 1000)

For a long time this method has been considered as the « gold standard » for toxin detection because of its very high sensitivity and specificity [6, 7]. Nevertheless it has some drawbacks. It lacks standardization and the technique varies from one laboratory to another according to the type of cell used, the dilution of the stool sample, the incubation period, etc. Furthermore, it requires a minimum of 24 to 48 hours incubation and facilities for cell culture. Lastly, the antiserum needed to confirm the specificity of the cytopathic effect is not available in France.*

Immuno-enzymatic methods

Over the last five years, many immuno-enzymatic tests have appeared on the French market [1, 2, 4, 5, 9-12, 14, 28, 32, 33, 38]. These kits detect either toxin A alone (Bartels™, Culturette toxin-CD™, Premier™, Tox A-test™, Vidas™) or both toxins simultaneously (Cytoclone™) (Table 2). They are for the most part sandwich type ELISA tests using two antitoxin antibodies, one of which is monoclonal, except for Bartels™ and Culturette™ toxin-CD, which have two polyclonal antibodies. The detection is colorimetric and is read either by direct vision or spectrophotometrically for all the kits except for Vidas™, which is fluorimetric. The sample is prepared by simple dilution of the stool, though the Bartels™ and Vidas™ kits require filtration or centrifugation of the stool suspension. These tests are very quick (< 3 hours) and very simple to perform with ready-to-use reagents, which makes them suitable for all laboratories. Only the Vidas™ method requires prior expenditure on specific automation.

Although all these tests have excellent specificity, they share two important disadvantages. The first is their lack of sensitivity as compared with SCA, which varies from 69% to 88% according to the kit used [2, 5, 9, 10, 12, 33]. The detection threshold lies between 10 and 1000 pg of toxin A.

When the tests are evaluated not only with SCA but also with regard to the clinical, endoscopic and bacteriological findings, wide variations in sensitivity are seen according to different studies (Table 3). These are due to the different clinical and laboratory criteria used in the diagnosis of C. difficile related diarrhea or colitis, to their prevalence in each study, to the method of storage of the stool specimen (fresh, or stored at +4°C or -70°C) and to the techniques used in carrying out the test, which vary with local practice and the complexity of the kit. Nonetheless published data seem to indicate two trends. The first is the relatively weak sensitivity of the Vidas™ test and its high incidence of uncertain results, and the second is the greater sensitivity of the Cytoclone™ test, probably due to the detection of toxins A and B and to the inclusion of the streptavidine-biotin system for amplifying the detection signal of the antigen-antibody complex.

Apart from the lack of sensitivity, the second disadvantage of these tests is their relatively high price, though this varies greatly between the manufacturers.

---

*Reagent kits (Bartels Immunodiagnostic C. difficile toxin™ Bellevue, Washington, USA; Cell-med C. difficile toxin™ Beldico, Belgium) are marketed in some countries, but often at a prohibitive price. They may nonetheless represent a possible approach towards standardization.

To complete the list of immuno-enzymatic methods, one should mention the existence of a rapid test (< 15 min) for the detection of toxin A provided in the form of individual cassettes (Diff-Cube™ Difco) [42]. Although it claims a 92% correlation with SCA, this kit has never been marketed in France. More recently, two other rapid tests (Immunocard™ toxin A, Meridian; *C. difficile* toxin A™, Oxoid) detecting toxin A in less than 40 minutes have been commercially available in France. Their clinical evaluation is currently under investigation.

Table 2. Commercial kits for the immuno-enzymatic detection of *C. difficile* toxins

Name	Manufacturer	Characteristics
Bartels™	Baxter	- polyclonal Ab anti-tox A - 2.5 hours - centrifugation of stool
Culturette™ toxin-CD	Becton Dickinson	- polyclonal Ab anti-tox A - 1.25 hours
Cytoclone™ A+B	Meridian	- monoclonal Ab anti-tox A + B - 1.50 hours - Streptavidine-Biotine
Tox A-test™	TechLab	- monoclonal Ab anti-tox A - 2.5 hours
Premier ™	Meridian	- monoclonal Ab anti-tox A - 1.5 hours
Vidas™	bioMérieux	- monoclonal Ab anti-tox A - 2.5 hours - fluorimetric detection - automatized method - centrifugation of stool

Ab = antibody

Table 3. ELISA tests for diagnosis of *C. difficile* infection. Review of the literature

Name	Sensitivity (%)	Specificity (%)	Indeterminate results (%)
Bartels™	87-93	91-96	2.4-3
Culturette™ toxin-CD	83-85	97-98	1-3
Cytoclone™ A+B	75-89	97-100	1.5-3.2
Tox A-test™	87-99	91-95	3-15
Premier™	65-93	87-100	0.3-2.1
Vidas™	52-80	75-100	2.5-19

Molecular biology

Cloning and gene sequencing of toxins A and B has opened the way to molecular biological methods for the diagnosis of *C. difficile* infection.

The hybridization of a stool suspension deposited on a nitrocellulose membrane with the aid of oligonucleotide probe of 33 pb. labelled with ^{32}P, specific for toxin B, has recently been assessed by Green et al [14]. This method presents a sensitivity and specificity of 100% and 83% respectively, as compared with SCA. It is quick (< 24 hours) and gives concrete results. Its drawback is the difficulty of using radio-isotopes. Labelling the probes with digoxigenin may provide an alternative, but it is difficult to read the signal because of a high background. A method of hybridizing the colonies by means of a probe specific for toxin B, labelled with digoxigenin has shown an excellent sensitivity of 100% as compared with the SCA, but a lesser specificity (96.7%) [40].

Polymerase chain reaction (PCR) has been used to amplify a genomic fragment coding for the rRNA 16S specific for *C. difficile*, as for toxins A and B [15, 16, 18, 19, 22, 27, 41, 43]. This method was first used for isolated strains of *C. difficile* before being applied directly to stool samples. Table 4 summarizes the type of probes used by different teams. The detection threshold of PCR is estimated as lying between 10 and 10^3 organisms per gram of stool. Some authors have reported the presence of substances such as bile salts, urea and bilirubin inhibit the amplification, forcing the bacteriologist to go through several steps in the purification of DNA. To avoid this, Wolfhagen et al [41] introduced a preliminary rapid extraction stage, by the use of magnetic spheres coated with an antibody specific for *C. difficile* (the Magnetic Immuno PCR Assay - MIPA). This method, which has already been applied to salmonella, is quick, simple and efficacious but involves a lowering of the detection threshold to 10^3-10^4 organisms per gram of stool. Whatever method of preliminary extraction/purification of DNA is used, the specificity of PCR for detecting the toxins of *C. difficile* is excellent, and there is no cross reaction with the hemorrhagic toxin of *C. sordellii*.

When applied to clinical samples, PCR will detect non-viable or poorly growing bacteria. Furthermore, it seems able to predict recurrences of *C. difficile* diarrhea at

Table 4. Detection of *C. difficile* and its toxins by PCR

Gene	Primers	Probe	Medium	Detection threshold (bact./g of stool)	References
Toxin B (399 pb)	YT-17 YT-18	YT-20 biotin-labeled	stools culture	10^2-10^3	Gumerlock, 1993 Kuhl, 1993
Toxin A (252 pb)	NK-2 NK-3	NK-12 ^{32}P	stools culture	?	Kato, 1993 Kato, 1991
16 S rRNA (270 pb)	B PG-48	PG-49 biotin-labeled	stools culture	10	Gumerlock, 1991
Toxins A+ B	Multiplex		culture	?	McMillin, 1992

an earlier stage than does stool culture or the SCA [22]. Finally, the simultaneous detection of genes coding for toxin B and for the rRNA 16S of *C. difficile* allows rapid distinction between the carriers of toxic and non-toxic strains.

In spite of these advantages, PCR is still far from being in routine laboratory use in the diagnosis of *C. difficile* infection. The DNA extraction process is long and difficult to perform, and requires special staff and materials.

Interpretation of the results

The detection of the toxins, the main factors determining the virulence of a strain of *C. difficile*, in the stools has for a long time been the benchmark method for the diagnosis of *C. difficile* related disease. However clinical experience has shown that the interpretation of a positive or negative result must be cautious, and always taken in the context of the clinical and epidemiological circumstances.

In fact, *C. difficile* toxins are often found in children of up to two years of age, who are asymptomatic [36]. Furthermore, the toxins may persist in patients suffering from *C. difficile* related diarrhea although a course of vancomycin or metronidazole seems to have been clinically successful [34].

On the other hand, although the production *in vitro* of toxins A and B is equal, it may be influenced *in vivo* by factors such as the diet, resulting in a lesser production of toxin A with no change in that of toxin B [24]. This observation may explain the lack of correlation between the levels of toxin B in the stools of patients with *C. difficile* infection, as measured by SCA and optical densities obtained by immuno-enzymatic methods detecting toxin A [5].

Finally, the direct detection of toxins in the stools does not have a 100% negative predictive value for the diagnosis of *C. difficile* related disease. Gerding et al [13] have shown that 11% of patients suspected of having the infection who were stool toxin-negative and toxinogenic culture-positive were found on endoscopy to have pseudomembranes.

These three points illustrate the difficulty of interpreting the microbiological results, and should stimulate the investigator to employ several diagnostic methods such as stool culture and toxin detection, and above all to correlate the findings with the clinical and endoscopic data.

Conclusion

The diagnosis of *C. difficile* infection has been made much easier over the last few years by the introduction of immuno-enzymatic tests. These have allowed many laboratories to detect the toxins of *C. difficile* independently, without recourse to a "specialist". The tests are simple, quick and highly specific. They lend themselves well to the emergency situation and represent an important step forward in the fight against nosocomial infections resulting from this organism. However, their sensitivity varies with the quality of the kit, and is still below that of the cytotoxicity test, which remains the gold standard for the detection of toxin B in the stools. As far as PCR is concerned, its complexity and lack of standardization make it difficult to apply in current practice.

References

1. Altaie SS, Meyer P, Dryja D (1994) Comparison of two commercially available enzyme immunoassays for detection of *Clostridium difficile* in stool specimens. J Clin Microbiol 32: 51-53

2. Barbut F, Caburet F, Petit JC (1992) Évaluation d'un test immunoenzymatique (ELISA) détectant la toxine A de *Clostridium difficile* dans les échantillons de selles. Ann Biol Clin 50 : 31-36

3. Barbut F, Corthier G, Petit JC (1992) Pathophysiology of *Clostridium difficile*-associated intestinal diseases. Médecine Sciences 3: 214-222

4. Barbut F, Kajzer C, Planas N, Petit JC (1993) Comparison of three enzyme immunoassays, a cytotoxicity assay, and toxigenic culture for diagnosis of *Clostridium difficile*-associated diarrhea. J Clin Microbiol 31: 963-967

5. Borriello SP, Vale T, Brazier JS, Hyde S, Chippeck E (1992) Evaluation of a commercial enzyme immunoassay kit for the detection of *Clostridium difficile* toxin A. Eur J Clin Microbiol Infect Dis 11: 360-363

6. Bowman RA, Riley TV (1988) Laboratory diagnosis of *Clostridium difficile*-associated diarrhoea. Eur J Clin Microbiol Infect Dis 7: 476-484

7. Brazier JS (1993) Role of the laboratory in investigations of *Clostridium difficile* diarrhea. Clin Infect Dis 16 (suppl 4): S228-S233

8. Chang TW, Lauermann M, Bartlett JG (1979) Cytotoxicity assay in antibiotic-associated colitis. J Infect Dis 140: 765-770

9. De Girolami PC, Hanff PA, Eidhelberger K, Longhi L, Teresa H, Pratt J, Cheng A, Letourneau JM, Thorne GM (1992) Multicenter evaluation of a new enzyme immunoassay for detection of *Clostridium difficile* enterotoxin A. J Clin Microbiol 30: 214-219

10. Delmée M, Mackey T, Amitou A (1992) Evaluation of a new commercial *Clostridium difficile* toxin A enzyme immunoassay using diarrhoeal stools. Eur J Clin Microbiol Infect Dis 11: 246-249

11. DiPersio JR, Varga FJ, Conwell DL, Kraft JA, Kozak KJ, Willis DH (1992) Development of a rapid enzyme immunoassay for *Clostridium difficile* toxin A and its use in the diagnosis of *C. difficile*- associated disease. J Clin Microbiol 29: 2724-2730

12. Doern GV, Coughlin RT, Wu L (1992) Laboratory diagnosis of *Clostridium difficile*-associated disease: comparison of a monoclonal antibody enzyme immunoassay for toxins A and B with a monoclonal antibody enzyme immunoassay for toxin A only and two cytotoxicity assays. J Clin Microbiol 30: 2042-2046

13. Gerding DN, Olson M, Peterson R, Teasley DG, Gebhard RL, Schartz ML, Lee JT Jr (1986) *Clostridium difficile*-associated diarrhea and colitis in adults. A prospective case controlled epidemiologic study. Arch Intern Med 146: 95-100

14. Green GA, Riot B, Monteil H (1994) Evaluation of an oligonucleotide probe and an immunological test for direct detection of toxigenic *Clostridium difficile* in stool samples. Eur J Clin Microbiol Infect Dis 13: 576-581

15. Gumerlock H, Tang YJ, Meyers FJ, Silva J (1991) Use of polymerase chain reaction for the specific and direct detection of *Clostridium difficile* in human feces. Rev Infect Dis 13: 1053-1060

16. Gumerlock H, Tang YJ, Weiss JB, Silva J (1993) Specific detection of toxigenic strains of *Clostridium difficile* in stool specimens. J Clin Microbiol 31: 507-511

17. Johnson S, Clabots CR, Linn FV, Olson MM, Peterson LR, Gerding DN (1990) Nosocomial *Clostridium difficile* colonization and disease. Lancet 336: 97-100

18. Kato N, Ou CY, Kato H, Bartley SL, Brown VK, Dowell VR, Ueno K (1991) Identification of toxigenic *Clostridium difficile* by the polymerase chain reaction. J Clin Microbiol 29: 33-37

19. Kato N, Ou CY, Kato H, Bartley SL, Luo CC, Killgore GE, Ueno K (1993) Detection of toxigenic *Clostridium difficile* in stool specimens by the polymerase chain reaction. J Infect Dis 167: 455-458

20. Kelly MT, Champagne SG, Sherlock CH, Noble MA, Freeman HJ, Smith JA (1987) Commercial latex agglutination test for detection of *Clostridium difficile*-associated diarrhea. J Clin Microbiol 25: 1244-1247

21. Kelly CP, Pothoulakis C, LaMont JT (1994) *Clostridium difficile* colitis. N Engl J Med 330: 257-262

22. Kuhl SJ, Tang YJ, Navarro L, Gumerlock PH, Silva J (1993) Diagnosis and monitoring of *Clostridium difficile* infections with the polymerase chain reaction. Clin Infect Dis 16(suppl 4): S234-S238

23. Lyerly DM, Wilkins TD (1986) Commercial latex for *Clostridium difficile* toxin A does not detect toxin A. J Clin Microbiol 23: 622-623

24. Mahe S, Corthier G, Dubos F (1987) Effects of various diets on toxin production by two strains of *Clostridium difficile* in gnotobiotic mice. Infect Immun 55: 1801-1805

25. Maniar AC, Williams JW, Hammond GW (1987) Detection of *Clostridium difficile* toxin in various tissue culture monolayers. J Clin Microbiol 25: 1999-2000

26. McFarland LV, Mulligan ME, Kwok RYY, Stamm WE (1989) Nosocomial acquisition of *Clostridium difficile* infection. N Engl J Med 320: 204-210

27. McMillin DE, Muldrow LL, Laggette SJ (1992) Simultaneous detection of toxin A and toxin B genetic determinants of *Clostridium difficile* using the multiplex polymerase chain reaction. Can J Microbiol 38: 81-83

28. Merz CS, Kramer C, Forman M, Gluck L, Mills K, Senft K, Steiman I, Wallace N, Charache P (1994) Comparison of four commercially available rapid enzyme immunoassays with cytotoxin assay for detection of *Clostridium difficile* toxin(s) from stool specimens. J Clin Microbiol 32: 1142-1147

29. Murray PR, Weber CJ (1983) Detection of *Clostridium difficile* cytotoxin in HeP2 and CHO cell lines. Diagn Microbiol Infect Dis 1: 331-333

30. Peterson LR, Olson MM, Shanholtzer CJ, Gerding CJ (1988) Results of a prospective, 18-month clinical evaluation of culture, cytotoxin testing, and culturette brand (CDT) latex testing in the diagnosis of *Clostridium difficile*-associated diarrhea. Diagn Microbiol Infect Dis 10: 85-90

31. Poxton IR, Byrne MD (1981) Detection of *Clostridium difficile* toxin by counter-immuno-electrophoresis: a note of caution. J Clin Microbiol 14: 349

32. Schué V, Green G A, Monteil H (1995) Comparison of the Tox-A test with the cytotoxicity assay and culture for the detection of *Clostridium difficile*-associated diarrhoeal disease. J Med Microbiol 41: 316-318

33. Shanholtzer CJ, Willard KE, Holter JJ, Olson MM, Gerding DN, Peterson LR (1992) Comparison of the VIDAS *Clostridium difficile* toxin A with *C. difficile* culture and cytotoxin and latex tests. J Clin Microbiol 30: 1837-1840

34. Teasley DG, Gerding DN, Olson MM, Peterson LR, Gerhard RL, Schwartz MJ, Lee JT Jr (1983) Prospective randomised trial of metronidazole versus vancomycin for *Clostridium difficile-* associated diarrhoea and colitis. Lancet ii: 1043-1046

35. Tichota-Lee J, Jaqua-Stewart MJ, Benfield D, Simmons JL, Jaqua RA (1987) Effect of age on the sensitivity of cell cultures to *Clostridium difficile* toxin. Diagn Microbiol Infect Dis 8: 203-214

36. Viscidi R, Willey S, Bartlett JG (1981) Isolation rates and toxigenic potential of *Clostridium difficile* isolates from various patient populations. Gastroenterology 81: 5-9

37. West SEH, Wilkins TD (1982) Problems associated with counterimmuno-electrophoresis assays for detecting *Clostridium difficile* toxin. J Clin Microbiol 15: 347-349

38. Whittier S, Shapiro DS, Kelly WF, Walden P, Wait P, McMillon LT, Gilligan PH (1993) Evaluation of four commercially available enzyme immunoassays for laboratory diagnosis of *Clostridium difficile*-associated diseases. J Clin Microbiol 31: 2861-2865

39. Wilkins TD (1987) Role of *Clostridium difficile* toxins in disease. Gastroenterology 93: 389-391

40. Wolfhagen MJ, Fluit AC, Jansze M, Rademaker KC, Verhoef J (1993) Detection of toxigenic *Clostridium difficile* in fecal samples by colony blot hybridization. Eur J Clin Microbiol Infect Dis 12: 463-466

41. Wolfhagen MJ, Fluit AC, Torensma R, Popellier MJ, Verhoef J (1994) Rapid detection of toxigenic *Clostridium difficile* in fecal samples by magnetic immuno PCR assay. J Clin Microbiol 32: 1629-1633

42. Woods GL, Iwen PC (1990) Comparison of a dot immunobinding assay, latex agglutination, and cytotoxin assay for laboratory diagnosis of *Clostridium difficile*-associated disease. J Clin Microbiol 28: 855-857

43. Wren BW, Clayton CL, Casteldine NG, Tabaqchali S (1990) Identification of toxigenic *Clostridium difficile* strains by using a toxin A gene-specific probe. J Clin Microbiol 28: 1808-1812

44. Wu TC, Fung JC (1983) Evaluation of the usefulness of counterimmuno-electrophoresis for diagnosis of *Clostridium difficile*-associated colitis in clinical specimens. J Clin Microbiol 17: 610-613

Updates on the management of
Clostridium difficile associated intestinal disease

Christina M. Surawicz

SUMMARY

The spectrum of diarrheal disease associated with antibiotic therapy ranges from antibiotic associated diarrhea to the more severe pseudomembranous colitis (PMC) which is always associated with *Clostridium difficile*.

The initial management of uncomplicated mild diarrhea associated with antibiotic therapy includes stopping the antibiotic if at all possible, and assuring that the patient does not become dehydrated from diarrhea. For more severe symptoms, and certainly for colitis (suggested by the presence of severe diarrhea or blood in the stools, fever and abdominal pain, and when possible confirmed by endoscopy and biopsies), the therapeutic choices include: vancomycin 125-500 mg four times a day for ten days or metronidazole 250 mg four times a day for ten days. Clinical trials indicate that the drugs are equally effective in mild to moderate disease and that vancomycin 125 mg four times a day is equivalent to 500 mg four times a day. Metronidazole is less expensive ($30 for a 10-day course) but does have some side effects including a metallic taste in the mouth, anorexia, an "antabuse" effect and rare but irreversible peripheral neuropathy. Vancomycin is more expensive ($200-500 for a 10-day course); it has little to no absorption from the gastrointestinal tract and side effects are thus uncommon. While metronidazole is a reasonable first choice drug, the author favors the use of vancomycin in patients who are ill with colitis or who have not improved on metronidazole therapy. Other oral antibiotics with efficacy against *Clostridium difficile* include bacitracin, fusidic acid, and the glycopeptide antibiotic teicoplanin. A difficult treatment dilemma in *Clostridium difficile* disease occurs when there is associated ileus or toxic colon. In this setting, patients are critically ill and delivery of the antibiotic to the colon is difficult. The therapeutic approach should involve IV metronidazole and vancomycin as well as vancomycin given by enemas and/or by nasogastric tube. Occasionally it may be necessary to deliver the vancomycin to the colonic lumen via a colonoscopically-placed decompression tube or via a surgical cecostomy. Another difficult decision is the role of surgery in critically ill patients. Surgery should be considered if there is worsening organ failure, severe colitis (documented by computerized tomographic abdominal scan or colonoscopy) which is refractory to therapy or if there is evidence of perforation or peritonitis. The surgical approach should be subtotal colectomy. It is hoped that early diagnosis and therapy will avoid the necessity of surgery in most cases.

Acquisitions récentes sur la prise en charge des pathologies intestinales associées à *Clostridium difficile*

RÉSUMÉ

Le spectre des pathologies diarrhéiques associées à l'antibiothérapie s'étend de la diarrhée simple associée aux antibiotiques, aux formes les plus sévères de colite pseudomembraneuse (CPM) qui sont toujours associées à *Clostridium difficile*.

Le traitement initial d'une diarrhée bénigne non compliquée associée à une antibiothérapie comprend l'arrêt de l'antibiotique lorsqu'il est possible, et le maintien d'une hydratation correcte du patient. Lorsque les symptômes sont plus sévères, et a fortiori s'il existe une colite, les thérapeutiques de choix seront la vancomycine, 125 à 500 mg quatre fois par jour pendant 10 jours, ou le métronidazole, 250 mg quatre fois par jour pendant 10 jours. Le diagnostic de colite est évoqué par la présence d'une diarrhée sévère et de sang dans les selles, d'une fièvre et de douleurs abdominales, et confirmé si possible par l'endoscopie et la réalisation de biopsies.

Les essais cliniques mettent en évidence une efficacité équivalente des deux médicaments dans les formes bénignes à modérées et, pour la vancomycine, des résultats identiques aux doses de 125 mg quatre fois par jour ou de 500 mg quatre fois par jour. Le coût du métronidazole est moins élevé (30 $ US pour un traitement de 10 jours), mais il entraîne un certain nombre d'effets secondaires tels qu'un goût métallique dans la bouche, une anorexie, un effet antabuse et, parfois, une neuropathie périphérique dont la survenue est rare mais irréversible. La vancomycine a un coût de revient plus élevé (200 à 500 $ US pour un traitement de 10 jours). Son absorption au niveau du tractus gastro-intestinal est minime ou inexistante, et les effets secondaires sont par conséquent très rares.

Le métronidazole constitue un choix logique pour un médicament de première intention, mais les auteurs recommandent l'usage de la vancomycine en cas de colite ou en l'absence de réponse au métronidazole. Les autres antibiotiques administrés per os et efficaces contre *C. difficile* sont la bacitracine, l'acide fusidique et la teicoplanine (antibiotique glycopeptidique). Dans les pathologies associées à *C. difficile*, la décision thérapeutique est particulièrement délicate quand sont présents un iléus ou un côlon toxique. Dans cette situation, les patients sont à un stade critique, et la diffusion de l'antibiotique au niveau du côlon est difficile. L'approche thérapeutique doit comprendre le métronidazole ou la vancomycine par voie intraveineuse, ou bien la vancomycine administrée par lavements ou par sonde nasogastrique.

Il est parfois nécessaire d'administrer la vancomycine directement dans la lumière colique par un tube de décompression placé par colonoscopie, ou par une cæcostomie chirurgicale. Une autre difficulté du choix thérapeutique est le rôle de la chirurgie chez les patients dont l'état est critique. La chirurgie doit être envisagée en cas d'aggravation d'une défaillance organique, s'il existe une colite sévère documentée par scanner abdominal ou par colonoscopie et réfractaire au traitement, ou devant une perforation ou une péritonite. La technique chirurgicale est alors la colectomie subtotale. On peut espérer qu'un diagnostic plus précoce et la mise en place d'un traitement adapté, permettront d'éviter dans la plupart des cas le recours à la chirurgie.

Neue Erkenntnisse zur Behandlung
von *Clostridium difficile*-assoziierten Darmerkrankungen

ZUSAMMENFASSUNG

Das Spektrum der Durchfallerkrankungen im Zusammenhang mit Antibiotika-therapien reicht von Antibiotika-assoziierten Diarrhöen bis zur ernsteren pseudo-membranösen Kolitis (PMC), an der stets *Clostridium difficile* beteiligt ist. Zur Behandlung unkomplizierter harmloser Diarrhöen im Zusammenhang mit einer Antibiotikatherapie gehören zunächst, falls irgend möglich, das Absetzen des Antibiotikums sowie Maßnahmen zur Verhütung einer Dehydratation des Patienten durch die Diarrhö. Bei schwererwiegenden Symptomen und natürlich bei Kolitis (erkennbar am Vorliegen starker Diarrhöen oder von Blut im Stuhl, Fieber und Bauchschmerzen und möglichst endoskopisch oder durch Biopsien gesichert), sind die Mittel der Wahl entweder zehn Tage lang viermal täglich 125-500 mg Vancomycin oder zehn Tage lang viermal täglich 250 mg Metronidazol. In klinischen Prüfungen zeigte sich, daß bei leichten bis mittelschweren Erkrankungen beide Substanzen gleichermaßen wirksam sind und daß Vancomycin in einer Dosierung von viermal täglich 125 mg genauso wirksam ist wie in einer Dosierung von viermal täglich 500 mg. Metronidazol ist preiswerter (eine zehntägige Behandlung kostet $30), hat jedoch einige Nebenwirkungen, darunter einen metallischen Geschmack im Mund, Anorexie, einen "Antabus"-Effekt sowie seltene aber irreversible Fälle peripherer Neuropathien. Vancomycin ist teurer (eine zehntägige Behandlung kostet $200-500); es wird nur in sehr geringem Umfang oder gar nicht aus dem Darm resorbiert, so daß Nebenwirkungen unüblich sind. Obwohl Metronidazol ein vernünftiges Mittel der Wahl ist, bevorzugt die Autorin Vancomycin bei Patienten, die an einer Kolitis leiden oder deren Zustand sich unter Metronidazol nicht gebessert hat. Andere gegen *Clostridium difficile* wirksame orale Antibiotika sind Bacitracin, Fusidinsäure und das Glykopeptid Teicoplanin. Ein schwieriges Dilemma bei der Behandlung von Erkrankungen im Zusammenhang mit *C. difficile* ist mit dem Vorliegen eines Ileus oder eines toxischen Megakolon verbunden. In diesen Fällen ist das Einbringen des Antibiotikums in den Dickdarm aufgrund des schlechten Gesundheitszustands der Patienten erschwert. Die Behandlung sollte mit Metronidazol und Vancomycin i.v. sowie Vancomycin in Form von Einläufen und/oder Nasenmagensonde erfolgen. Gelegentlich muß zur Verabreichung des Vancomycins koloskopisch ein Unterdruckschlauch in das Dickdarmlumen eingebracht oder chirurgisch ein Zäkostoma angelegt werden. Ebenfalls problematisch ist die Entscheidung über chirurgische Eingriffe bei lebensgefährlich erkrankten Patienten. Solche Eingriffe sollten in Betracht gezogen werden bei zunehmendem Organversagen, therapieresistenter schwerer Kolitis (durch CT des Abdomens oder Koloskopie nachgewiesen) oder Anhaltspunkten für eine Perforation oder Peritonitis. Chirurgisch sollte eine subtotale Kolonresektion erfolgen. Es steht zu hoffen, daß durch Früherkennung und -behandlung der Notwendigkeit chirurgischer Eingriffe in der Mehrheit der Fälle vorgebeugt werden kann.

Introduction

Clostridium difficile is the most common nosocomial infection of the gastro-intestinal tract. Epidemics have been documented in hospital settings as well as in nursing homes and rehabilitation centers [22]. The association of *C. difficile* disease with antibiotic therapy is well recognized.

Many newer broad spectrum antibiotics may predispose to its acquisition. *C. difficile* can also cause diarrhea in previously healthy individuals who are given antibiotics in an out-patient setting. Several excellent clinical reviews have recently been published [4,11,17]. This paper will focus on therapy for *Clostridium difficile* associated disease and as well as illness complicated by ileus or toxic colon.

Therapy

Table 1 summarizes the proposed treatment for *Clostridium difficile* disease.

Antibiotic associated diarrhea (AAD)

Diarrhea is a frequent side effect of antibiotics, occurring in estimates of up to 20% of individuals who receive them. *C. difficile* is found in the stools in only 20% - 30%, suggesting that other factors are likely contributing to the diarrhea. One such factor may be the alteration in normal fecal flora, with resultant changes in colonic carbohydrate fermentation and short chain fatty acid production which may contribute to an osmotic diarrhea and/or to a decrease of sodium absorption normally stimulated by short chain fatty acids.

It is important that patients with diarrhea remain well hydrated to counteract fluid loss and volume depletion. A reasonable initial approach is to discontinue antibiotics, if possible; this should allow the re-establishment of the normal colonic microflora. If this is not feasible, one may consider changing the antibiotic to one with a more narrow spectrum, especially if it can allow re-population of the colonic anaerobic bacteria which are important in maintaining colonization resistance.

The use of antidiarrheal agents in AAD is controversial. However, since there is evidence that they should be avoided in PMC [5], it is probably prudent to avoid their use in AAD as well.

Pseudomembranous colitis (PMC)

C. difficile associated diarrhea should be suspected in anyone who develops diarrhea during antibiotic therapy. Diarrhea can even occur up to 8 weeks after the end of a course of antibiotics. It is more common after oral antibiotics, and has even been reported as a complication of single dose cephalosporins given pre-operatively. The diagnosis is made by detection of *C. difficile* and/or cytotoxin B in the stool [21]. Colitis should be suspected when there is fever, abdominal pain, and diarrhea with gross or occult blood in the stools. Diagnosis of PMC can be definitively determined only when pseudomembranes are observed with a sigmoidoscopic or colonoscopic examination. White blood cells may be present in the stools but are not a reliable indicator of colitis.

Table 1. Treatment of *Clostridium difficile* disease

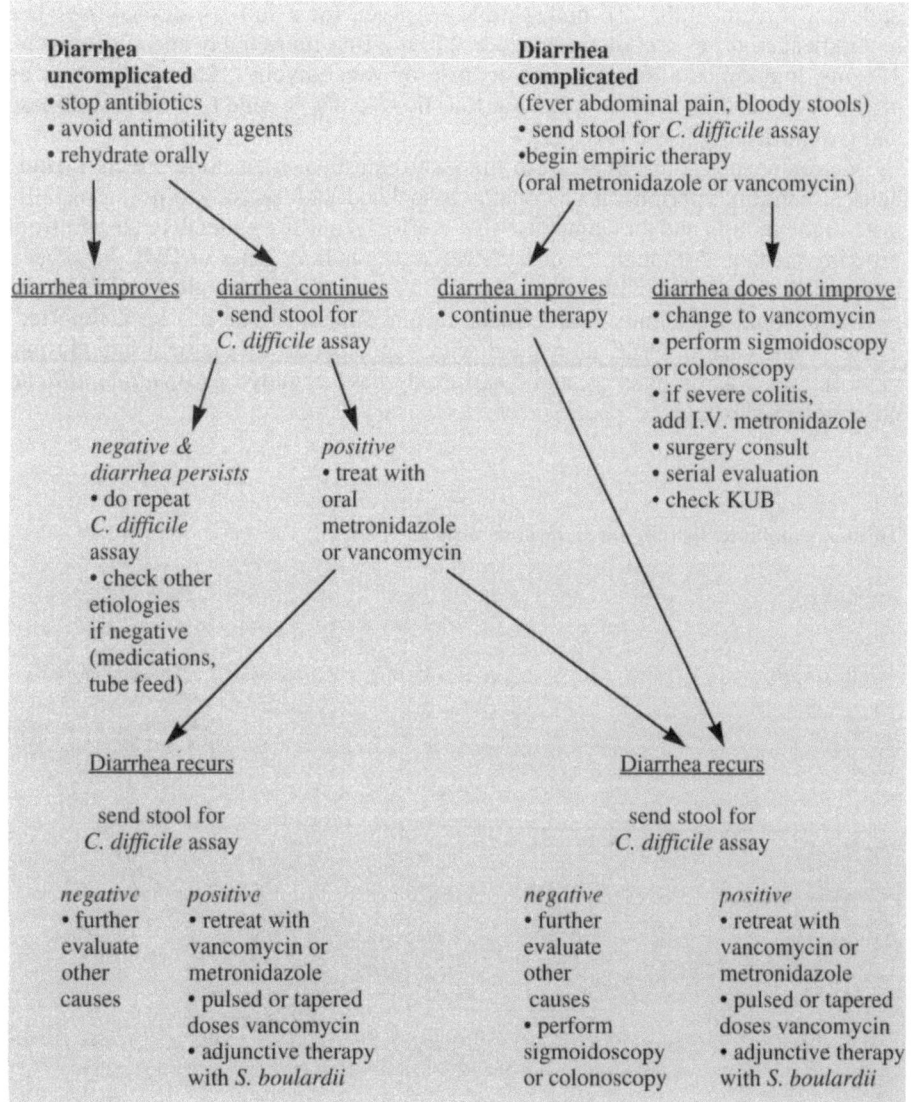

They were absent in 72% of toxin positive stools in one study. In cases of colitis, *C. difficile* culture and toxin will be positive in 90 - 100% of cases.

Initial evaluation should assess for hydration and severity of illness. The diagnosis rests on stool examination, but results will not be available for 24 - 48 hours, thus empiric therapy is often necessary. It is reasonable to treat for presumptive *C. difficile* in anyone who develops diarrhea in the hospital, in some cases of chronic diarrhea, when symptoms worsen or progress, when colitis is present, or in anyone who has a history of prior *C. difficile* infection.

Vancomycin was the first agent used to treat PMC [26, 29]. Currently the two most commonly used antibiotics used to treat *C. difficile* disease are metronidazole and vancomycin (Table 2); both should be given for a full ten day course. The recommended doses are: metronidazole, 250 mg four times a day and vancomycin, 125 mg four times a day. The lower dose of vancomycin (125 mg qid) was as effective as the higher dose of 500 mg four times a day in mild to moderate disease and is significantly less costly [12].

Metronidazole is absorbed from the gastrointestinal tract. Side effects include nausea, vomiting, peripheral neuropathy associated with prolonged use, a metallic taste in the mouth, and an « antabuse-like » effect requiring patients to abstain from drinking alcohol. Although, *C. difficile* resistance is rare, cases of PMC have been reported in association with metronidazole. Vancomycin is not absorbed from the gastrointestinal tract, thus, side effects are uncommon. Rash has been reported. There is one case of PMC associated with vancomycin. The patient had chronic renal failure, was on dialysis, and could easily have acquired *Clostridium difficile* nosocomially [14].

Table 2. Antibiotic therapy for *C. difficile* disease

Antibiotic	Dose	Cost (10 day course)	Side Effects
Metronidazole	250 mg qid	$25 - 30	nausea, vomiting, diarrhea, metallic taste, «antabuse» effect, peripheral neuropathy (irreversible)
Vancomycin	125 mg qid	$300	rash (rare)
Bacitracin	25, 000 units qid	$150	bitter taste
Fusidic acid	0.5 - 1.5 g/day	Not available in USA	
Teicoplanin	100 mg bid	Not available in USA	

Clinical trials indicate that the two drugs are equivalent for the treatment of mild disease [28]. Thus, it is reasonable to use metronidazole as a first line drug. However, there are no trials comparing therapy in sicker patients. The author favors the use of vancomycin, when there is clinical evidence of colitis (abdominal pain, fever or blood in the stools), when colonoscopy reveals pseudomembranes, when the patient is sick, or when patient symptoms have not improved or resolved with metronidazole or another drug.

Other antibiotics are effective for *C. difficile* disease. Oral bacitracin has been given 80,000 units per day. However, it is expensive, and its use is limited by an unpalatable taste [30]. In one study it was less effective than vancomycin [10]. Fusidic acid has also been used to treat *C. difficile* [7]; it is not available in the US.

The glycopeptide antibiotic teicoplanin, given 100 mg twice a day for 10 days was shown to be equivalent in efficacy to vancomycin 500 mg four times a day [8, 9]. It may soon be available in the US. Other glycopeptides under investigation include eremomycin (studies in hamsters) and ramoplanin.

Nonantimicrobial therapy

The use of antidiarrheals in PMC is controversial, but there is suggestive evidence that they should be avoided in PMC. Agents such as diphenoxylate hydrochloride or loperamide do decrease diarrhea, but there is evidence that decreased transit may lengthen the duration of illness and lead to complications [5].

The bile salt binding resin cholestyramine has been used; this agent does cause constipation and might be helpful in decreasing diarrhea. The rationale for this agent came from the notion that the resin might bind bile salts, though this has not been supported by laboratory evaluation. However, resin binding of antibiotics could decrease their efficacy.

The nonpathogenic yeast *Saccharomyces boulardii* has been shown to be effective in decreasing the incidence of AAD in hospitalized patients receiving antibiotics [27]. When used as an adjunct to antibiotic therapy in first infections with *C. difficile* disease, it had no benefit over placebo in preventing recurrent disease. However, it was significantly more effective than placebo as an adjunct to antibiotic therapy in treatment of recurrent *C. difficile* disease in those who had already had one or more recurrence [23].

Clinical response

The diarrhea usually improves in one to four days, with resolution by two weeks. Most patients experiencing their first episode of *C. difficile* disease respond well to antibiotic therapy, but 12-24% develop further recurrence. Once patients have multiple episodes, the frequency of further recurrences increase [28] (Fig. 1).

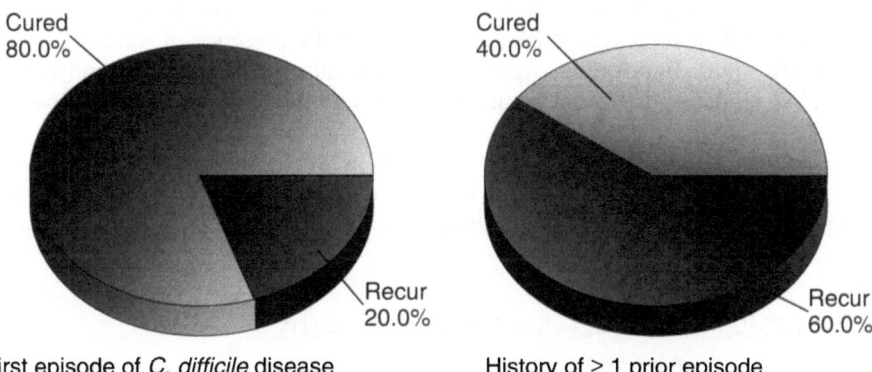

Cured 80.0%   Recur 20.0%   First episode of *C. difficile* disease

Cured 40.0%   Recur 60.0%   History of ≥ 1 prior episode of *C. difficile* disease

Fig. 1. Response to antibiotic treatment by history of *Clostridium difficile* disease. (From McFarland et al, JAMA, 1994)

Treatment when ileus or toxic colon is present

The patient with PMC who has an associated ileus or toxic colon or megacolon presents a therapeutic dilemma (Table 3) because it is difficult to deliver the oral antibiotics to the colon. Therapy should include intravenous metronidazole which penetrates intestinal tissue. The dose is 500 mg every 6-8 hours [18, 24]. Some feel that intravenous vancomycin may have no role, as failure may occur [2, 13]. The author, however, favors the advice of Fekety and Shah [11] who recommend use of parenteral vancomycin as well as metronidazole. This is such a desperate clinical situation that all measures should be tried to reverse the disease process.

Vancomycin can be given by nasogastric tube, or rectally as enemas. The author has had experience using a colonoscope to decompress a dilated colon, and then placing a colon decompression tube to relieve distension as well as to deliver vancomycin. On rare occasions, a surgical or endoscopic cecostomy may be necessary to decompress the colon and deliver vancomycin to the colonic lumen. In these critically ill patients, serial clinical evaluation is important to look for signs that urgent surgery may be needed. Mortality from fulminant toxic colitis or perforation ranges from 2 to 8%.

Table 3. Treatment of severe PMC (ileus, toxic colon)

• IV Metronidazole 500 mg q 6 hr	• PO Vancomycin per NG tube Enemas (500 mg/l) Cecostomy or tube
• IV Vancomycin. Keep serum 35 mg/ml	• Monitor closely

Indications for surgery

The main indication for surgery (Table 4) is worsening clinical condition despite adequate therapy. Clinical clues are organ failure, peritonitis (suggesting perforation), and progressive colitis. Progressive disease can be documented by abdominal CT scan, a very thick colon wall and the presence of ascites are poor prognostic signs. In a recent surgical series, patients with a worse prognosis were older, had an ileus, history of recent antibiotics, and ascites or very thick colon wall on CT scan [25].

The procedure of choice is subtotal colectomy, since segmental resection colitis often leads to re-operation to remove remaining diseased bowel [20]. The key to avoiding this grim situation is early diagnosis and therapy.

Table 4. PMC with toxic megacolon. Indications for surgery

• Increasing abdominal pain and distension	
• Subserosal air - KUB	• Vancomycin irrigation post-op
• Subtotal colectomy	• Last resort

Special situations

C. difficile culture-positive toxin negative diarrhea

Some individuals are *C. difficile* culture positive but toxin negative, with no evidence of colitis; in many, diarrhea will improve with therapy to eradicate *C. difficile* [19].

C. difficile colitis in persons with acquired immune deficiency syndrome (AIDS)

Patients with HIV infection acquire *C difficile* because of their frequent use of antibiotics. Studies suggest that they respond to therapy as well as non AIDS groups [3, 6]. Profound diarrhea can occur [3], but response to antibiotics is usually excellent.

C. difficile associated with cancer chemotherapy

Many chemotherapeutic agents predispose patients to acquisition of *C. difficile*. Recognition may be delayed and the problem may not be recognized as many of these people are also on antibiotics [1]. Prompt recognition and early therapy are important.

Prevention of transmission

Staff caring for patients with *C. difficile* infection in the hospital setting can prevent transmission by use of vinyl gloves between patients [15] and hand washing [22]. Asymptomatic carriers exist, but therapy is not recommended as treatment with vancomycin resulted in higher carrier rates [16].

References

1. Anand A, Glatt AE (1993) *Clostridium difficile* infection associated with antineoplastic chemotherapy: a review. Clinical Infectious Diseases 17:109-113
2. Bolton RP, Culshaw MA (1986) Faecal metronidazole concentrations during oral and intravenous therapy for antibiotic associated colitis due to *Clostridium difficle*. Gut 27:1169-1172
3. Cappell MS, Philogene C (1993) *Clostridium difficile* infection is a treatable cause of diarrhea in patients with advanced human immunodeficiency virus infection: a study of seven consecutive patients admitted from 1986 to 1992 to a universtiy teaching hospital. Am J Gastroenterol 88: 891-897
4. Caputo GM, Weitekamp MR, Bacon AE, Whitener C (1994) *Clostridium difficile* infection: a common clinical problem for the general internist. J Gen Intern Med 9:528-533
5. Church JM, Fazio VW (1986) A role for colonic stasis in the pathogenesis of disease related to *Clostridium difficile*. Dis Colon Rectum 29:804-809

6. Cozart JC, Kalangi SS, Clench MH, Taylor DR, Borucki MJ, Pollard RB, Soloway RD (1993) *Clostridium difficile* diarrhea in patients with AIDS versus non-AIDS controls. Methods of treatment and clinical response to treatment. Gastroenterol 16: 192-194

7. Cronberg S, Castor B, Thoren A (1984) Fusidic acid for the treatment antibiotic-associated colitis induced by *Clostridium difficile*. Infection 12: 276-279

8. de Lalla F, Nicolin R, Rinaldi E, Scarpelini P, Rigoli R, Manfrin V, Tramarin A (1992) Prospective study of oral teicoplanin versus oral vancomycin for therapy of pseudo-membranous colitis and *Clostridium difficile* - associated diarrhea. Antimicrob Agents Chemother 36: 2192-2195

9. de Lalla F, Privitera G, Rinaldi E, Ortisi G, Santoro D, Rizzardini G (1989) Treatment of *Clostridium difficile*-associated disease with teicoplanin. Antimicrob Agents Chemother 33: 1125-1127

10. Dudley MN, McLaughlin JC, Carrington G, et al (1986) Oral bacitracin vs. vancomycin therapy for *Clostridium difficile* - induced diarrhea: a randomized double-blind trial. Arch Intern Med 146: 1101-1104

11. Fekety R, Shah AB (1993) Diagnosis and treatment of *Clostridium difficile* colitis. JAMA 269: 71-51

12. Fekety R, Silva J, Kauffman C, Buggy B, Derry HG (1989)Treatment of antibiotic-associated *Clostridium difficile* colitis with oral vancomycin: comparison of two dosage regimens. Am J Med 86: 15-19

13. Guzman R, Kirkpatrick J, Forward K, Lim F (1988) Failure of parenteral metro-nidazole in the treatment of pseudomembranous colitis. J Infect Dis 158: 1146-1147

14. Hecht JR, Olinger EJ (1989) *Clostridium difficile* colitis secondary to intravenous vancomycin. Dig Dis Sci 34: 148-149

15. Johnson S, Gerding DN, Olson MM, Weiler MD, Hughes RA, Clabots CR, Peterson LR (1990) Prospective, controlled study of vinyl glove use to interrupt *Clostridium difficile* nosocomial transmission. Am J Med 88: 137-140

16. Johnson S, Homann SR, Bettin KM, Quick JN, Clabots CR, Peterson LR, Gerding DN (1992) Treatment of asymptomatic *Clostridium difficile* carriers (fecal excretors) with vancomycin or metronidazole. Ann Intern Med 117: 297-302

17. Kelly CP, LaMont JT (1994) *Clostridium difficile* colitis. N Engl J Med 330: 256-262

18. Kleinfeld DI, Sharpe RJ, Donta ST (1988) Parenteral therapy for antibiotic-associated pseudomembranous colitis (letter). J Infect Dis 157: 389

19. Lashner BA, Todorczuk J, Sahm DF, Hanauer SB (1986) *Clostridium difficile* culture-positive toxin-negative diarrhea. Am J Clin Gastroenterol 81: 940-943

20. Lipsett PA, Samanatarau OK, Tam ML, Bartlett JG, Lillemoe KD (1994) Pseudomembranous colitis: a surgical disease? Surgery 116: 491-496

21. Lyerly DM, Krivan HC, Wilkins TD (1988) *Clostridium difficile*: its disease and toxins. Clin Microbiol Rev 1: 1-18

22. McFarland LV, Mulligan ME, Kwok RYY, Stamm WE (1989) Nosocomial acquisition of *Clostridium difficile* infection. N Engl J Med 320: 204-210

23. McFarland LV, Surawicz CM, Greenberg RN, Fekety R, Elmer GW, Moyer KA, et al (1994) A randomized placebo-controlled trial of *Saccharomyces boulardii* in combination with standard antibiotics for *Clostridium difficile* disease. JAMA 271: 1913-1918

24. Oliva SL, Guglielmo BJ, Jacabs R, Pons VG (1989) Failure of intravenous vancomycin and intravenous metronidazole to prevent or treat antibiotic-associated pseudomembranous colitis. J Infect Dis 159: 1154-1155

25. Prendergast TM, Marini CP, D'Angelo AS, Sher ME, Cohen JR (1994) Surgical patients with pseudomembranous colitis: factors affecting progress. Surgery 116: 768-774

26. Silva J Jr, Batts DH, Fekety R, Plouffe JF, Rifkin GD, Baird L (1981) Treatment of *Clostridium difficile* colitis and diarrhea with vancomycin. Am J Med 71: 815-821

27. Surawicz CM, Elmer GW, Speelman P, et al (1989) Prevention of antibiotic-associated diarrhea by *Saccharomyces boulardii*: A prospective study. Gastroenterology 96: 981-988

28. Teasley DG, Gerding DN, Olson MM, Peterson LR, Gebhard RL, Schwartz MJ, Lee JT Jr (1983) Prospective randomized trial of metronidazole versus vancomycin for *Clostridium difficile* associated diarrhoea and colitis. Lancet ii: 1043-1046

29. Tedesco F, Markham R, Gurwith M, Christie D, Bartlett JG (1978) Oral vancomycin for antibotic-associated pseudomembranous colitis. Lancet i: 226-228

30. Young GP, Ward PB, Bayley N, et al (1985) Antibiotic-associated colitis due to *Clostridium difficile* : double-blind comparison of vancomycin with bacitracin. Gastroenterology 85: 1038-1045

Treatment of the recurrent *Clostridium difficile* diarrheal syndrome with *Saccharomyces boulardii*

Robert Fekety

SUMMARY

Clostridium difficile is the most frequent cause of antibiotic-associated diarrhea and colitis in hospitals. Although the illness generally responds well to treatment with oral vancomycin or metronidazole, approximately 20% of patients experience a recurrence within 6-8 weeks after the discontinuation of specific antibiotic therapy. Some unfortunate patients experience repeated episodes over a period of 6-12 months or more, and until recently there has been no scientifically proven way to terminate this syndrome, which is thought to be caused by poorly understood antibiotic-induced aberrations in the ecology of the fecal flora that normally inhibits the growth and toxin production of *C. difficile*. *Saccharomyces boulardii* (*Sb*), a yeast which is found naturally on some fruits, was reported in two studies to effectively prevent antibiotic associated diarrhea and was thus investigated as an effective treatment for *C. difficile* disease. A multicenter randomized placebo controlled trial of *S. boulardii* in prevention of recurrent CDAD was conducted at the Universities of Washington, Kentucky and Michigan, and the results were reported in 1994. It was designed to determine the efficacy of orally administered *S. boulardii* (two 250 mg capsules twice per day containing 3×10^{10} CFU/day of *Sb*) in prevention of a recurrence of CDAD in patients with either their first or a recurrent episode of *Clostridium difficile* diarrhea or colitis. Immunosuppressed patients were excluded from the study. Each of 124 participants was treated with either vancomycin or metronidazole, and then with *Sb* or a placebo as well, for at least the last four days of antibiotic therapy, after which *Sb* or placebo was continued for a total of four weeks. Patients were followed for an additional four weeks and suspect recurrences were confirmed by stool cultures and toxin tests. Among 64 patients with their first episode of CDAD, the rate of recurrence was 19.3% in patients receiving *Sb* and 24.2% with placebo (NSS). In 60 patients with recurrent CDAD, the rate of another recurrence was significantly lower with *Sb* (9 of 26, 34.6%) than it was among placebo recipients (22 of 34, 64.7%) (p = 0.04). The rates of recurrence were not related to whether vancomyin or metronidazole was used in treatment. There were no side effects of the use of *Sb* other than thirst and constipation.

In summary, *Saccharomyces boulardii* was found to be safe and significantly more effective than placebo in prevention of *C. difficile* associated diarrhea or colitis. The ability of *S. boulardii* to prevent CDAD in patients suffering from recurrent episodes is particulary interesting, as this is the only therapy that has been shown in a well-controlled study to be effective in terminating this troublesome syndrome. *Sb* is not pathogenic, it is not a component of the normal fecal flora, and it does not permanently colonize the colon. As yet, all the mechanism(s) of the protective effect of *Sb* have not been established. Further studies of *Sb* with different types of patients with *C. difficile* diarrheal syndromes are underway.

Traitement des infections récidivantes à *Clostridium difficile* par *Saccharomyces boulardii*

RÉSUMÉ

C. difficile est la cause la plus fréquente de diarrhée associée aux antibiotiques et de colites dans les hôpitaux. Bien que ces pathologies répondent généralement bien au traitement par vancomycine ou métronidazole per os, environ 20 % des patients présentent une récidive dans les 6 à 8 semaines après l'arrêt du traitement anti-biotique spécifique. Certains sont parfois atteints de récidives multiples, sur des périodes de 6 à 12 mois ou même davantage. Jusqu'à récemment, il n'y avait pas de moyen scientifiquement établi, pour juguler ce syndrome. Il serait provoqué par un déséquilibre, encore mal élucidé, de l'écologie de la flore intestinale par les antibiotiques. Cette flore inhibe normalement la croissance de *C. difficile* et la production de toxines. *S. boulardii* est une levure que l'on trouve naturellement sur certains fruits, et qui a été décrite dans deux études comme capable de prévenir efficacement les diarrhées associées aux antibiotiques. Dès lors, *S. boulardii* a fait l'objet de travaux pour confirmer son intérêt dans le traitement des pathologies associées à *C. difficile*. Une étude multicentrique randomisée et contrôlée contre placebo, évaluant l'efficacité de *S. boulardii* dans la prévention des formes récidivantes de colites et diarrhées associées à *C. difficile* (CDCD) a été réalisée dans les universités de Washington, de Kentucky et du Michigan. Les résultats ont été publiés en 1994. L'objectif était de déterminer l'efficacité de *S. boulardii* (administrée per os à la dose de 2 capsules de 250 mg deux fois par jour, soit 3×10^{10} CFU/jour de *S. boulardii*),dans la prévention des récidives de CDCD chez des patients présentant soit un premier épisode, soit une récidive de diarrhée ou colite associée à *C. difficile*. Les patients immunodéprimés ont été exclus de l'essai. Chacun des 124 participants était traité par vancomycine ou métronidazole, puis par *S. boulardii* ou un placebo pendant au moins les quatre derniers jours de l'antibiothérapie. Après l'arrêt du traitement antibiotique, *S. boulardii* ou le placebo était poursuivi pendant un total de 4 semaines. Les patients étaient encore suivis pendant 4 semaines supplémentaires, et les suspicions de récidives étaient confirmées par coproculture et tests de recherche des toxines. Parmi 64 patients présentant un premier épisode de CDCD, le taux de rechute a été de 19,3 % dans le groupe traité par *S. boulardii*, et de 24,2 % dans le groupe recevant le placebo (différence non significative). Sur 60 patients atteints de formes récidivantes de CDCD, la fréquence d'une nouvelle récidive a été significativement plus basse avec *S. boulardii* (9 sur 26 soit 34,6 %), qu'avec le placebo (22 sur 34, soit 64,7 %) (p = 0,04). La fréquence de survenue des récidives n'est pas liée à l'utilisation de la vancomycine ou du métronidazole pour le traitement. Il n'y a pas eu d'effet secondaire au cours de l'administration de *S. boulardii*, hormis quelques cas de soif et de constipation.

Au total, *S. boulardii* possède une grande sécurité d'emploi et une efficacité significativement supérieure à celle d'un placebo dans la prévention des diarrhées ou colites associées à *C. difficile*. L'efficacité de *S. boulardii* dans la prévention des CDCD chez les patients atteints de formes récidivantes est particulièrement intéressante, puisque *S. boulardii* représente la seule thérapeutique dont on ait pu démontrer, dans le cadre rigoureux d'une étude contrôlée, la capacité à juguler ce syndrome pénible et invalidant. *S. boulardii* n'est pas pathogène, il n'est pas un composant de la flore fécale normale et ne colonise pas le côlon. À l'heure actuelle, tous les mécanismes expliquant l'effet protecteur de *S. boulardii* n'ont pas été précisés. De nouveaux travaux évaluant l'administration de *S. boulardii* chez des patients présentant différents types de syndromes diarrhéiques associés à *C. difficile* sont en cours de réalisation.

Behandlung rezidivierender *Clostridium difficile*-Infektionen mit *Saccharomyces boulardii*

ZUSAMMENFASSUNG

Clostridium difficile ist der häufigste Verursacher Antibiotika-assoziierter Diarrhö und Kolitis im Krankenhaus. Auch wenn die Krankheit im allgemeinen gut auf orale Gaben von Vancomycin oder Metronidazol anspricht, kommt es bei rund 20 % der Patienten innerhalb von 6-8 Wochen nach dem Absetzen der spezifischen Antibiotikabehandlung zu Rezidiven. Manche unglückliche Patienten erleiden über einen Zeitraum von 6-12 Monaten oder sogar noch länger immer wieder solche Rezidive. Bis vor kurzem gab es keine wissenschaftlich belegte Methode, dieses Syndrom zu unterbinden, das vermutlich auf noch unzureichend geklärten antibiotikabedingten Entgleisungen der Darmflora beruht, die normalerweise das Wachstum und die Toxinproduktion von *C. difficile* verhindert. *Saccharomyces boulardii* (*Sb*), eine in der Natur auf bestimmten Früchten vorkommende Hefe, erwies sich in zwei Studien als wirksames Mittel zur Verhütung Antibiotika-assoziierter Diarrhöen und wurde deshalb im Hinblick auf eine Wirksamkeit gegen *C. difficile*-Erkrankungen geprüft. Eine randomisierte, Placebo-kontrollierte Multicenterstudie mit *Sb* zur Prävention rezidivierender CDAD wurde an den Universitäten von Washington, Kentucky und Michigan durchgeführt und die Ergebnisse 1994 berichtet. Das Studiendesign sah vor, die Wirksamkeit oraler Gaben von *S. boulardii* (zweimal täglich zwei Kapseln à 250 mg *Sb*, je zu 3×10^{10} koloniebildende Einheiten) zur Prävention von CDAD-Rezidiven bei Patienten zu prüfen, die entweder zum ersten Mal oder zum wiederholten Mal ein Rezidiv einer *C. difficile*-Diarrhö oder Kolitis erlitten. Immungeschwächte Patienten waren aus der Studie ausgeschlossen. Jeder der 124 Teilnehmer wurde entweder mit Vancomycin oder Metronidazol behandelt und dann zusätzlich mit *Sb* oder Placebo, und zwar mindestens an den letzten vier Tagen der Antibiotikatherapie, danach wurde die Einnahme von *Sb* bzw. Placebo weitere vier Wochen lang fortgesetzt. Die Patienten wurden weitere vier Wochen nachbeobachtet, bei Verdacht auf ein Rezidiv wurden Stuhlkulturen angelegt und auf Toxine untersucht. Von den 64 Patienten, die zum ersten Mal ein CDAD-Rezidiv erlitten, betrug die Rezidivrate bei denjenigen, die *Sb* erhalten hatten, 19,3 % gegenüber 24,2 % in der Placebogruppe (NS). Von den 60 Patienten mit rezidivierender CDAD lag die Rate weiterer Rezidive bei den Patienten unter *Sb* erheblich niedriger (9 von 26 = 34,6 %) als bei der Placebogruppe (22 von 34 = 64,7 %) (p = 0,04). Ob der Patient Vancomycin oder Metronidazol erhalten hatte, wirkte sich nicht auf die Rezidivrate aus. Außer Durst und Verstopfung wurden keinerlei Nebenwirkungen der Behandlung mit *Sb* berichtet.

Zusammenfassend erwies sich *Saccharomyces boulardii* als sicher und statistisch signifikant wirksamer als Placebo in der Verhütung von *C. difficile*-assoziierter Diarrhö oder Kolitis. Besonders interessant ist die Tatsache, daß *Sb* in der Lage war, CDAD bei Patienten mit rezidivierenden Episoden zu verhüten, denn dies ist die einzige Behandlungsform, für die in einer korrekt durchgeführten placebo-kontrollierten Studie eine Wirksamkeit gegen dieses problematische Syndrom bewiesen werden konnte. *Sb* ist apathogen, gehört nicht zur normalen Darmflora und besiedelt den Dickdarm nicht auf Dauer. Bisher sind noch nicht alle Mechanismen erforscht, auf denen die Schutzwirkung von *Sb* beruht. Weitere Studien zu *Sb* an unterschiedlichen Patienten mit *C. difficile*-bedingten Diarrhösyndromen laufen derzeit noch.

Introduction

As many as 5-10 percent of hospitalized patients who receive antibiotics develop diarrhea. *Clostridium difficile* accounts for 10-15 percent of all antibiotic associated diarrhea (AAD) and it is the cause of almost all antibiotic associated colitis (AAC), which can present as either pseudomembranous colitis (PMC) or nonspecific colitis, or mild nonspecific diarrhea. The antibiotic associated diarrhea (AAD) that is unrelated to *C. difficile* is thought to be caused by poorly understood changes in the normal enteric flora, is relatively benign, self-limited and responsive to simple supportive therapy. However, it is often responsible for prolongation of hospitalization and higher medical costs, is associated with an increased risk of developing nosocomial infection and a three-fold increase in mortality [13]. *C. difficile* associated disease (CDAD) is more serious, and ranges from severe pseudomembranous colitis to non-specific colitis and to simple diarrhea without proven colitis. CDAD has been steadily increasing in frequency since 1985, and it is now the fourth most frequent type of nosocomial infection reported in the Center for Disease Control's National Nosocomial Infection Survey (NNIS) [W Jarvis, personal communication]

The highest rates of CDAD are in adult medical and surgical patients, and clusters of the disease have been documented in many different types of hospitals and nursing homes.

Although each episode of CDAD generally responds well to treatment with oral vancomycin or metronidazole, *C. difficile* is not necessarily eliminated from the patient's stools by specific antibiotic therapy. Approximately 20 percent of patients experience one or more recurrences or relapses of CDAD, which usually begin within 2-8 weeks after the discontinuation of specific antibiotic therapy for CDAD. Some unfortunate patients experience repeated episodes of CDAD over a period of 1-2 years or longer. Once patients have had one recurrence they tend to have more, no matter what antibiotic was used to treat them.

Each episode usually responds promptly to therapy with vancomycin or metronidazole. However, until recently there has been no proven way to prevent further relapses or recurrences, which may be caused either by persistence of *C. difficile* or by reinfection with it. Persistence and reinfection are thought to be possible because of the permissive effects of still poorly understood antibiotic-induced perturbations in the ecology and colonization-resistance of the fecal flora, which is believed normally to inhibit the colonization, growth and toxin production of *C. difficile*.

Saccharomyces boulardii

The use of the live yeast *Saccharomyces boulardii* (which is found naturally on lychee fruit and has an unusual optimum growth temperature of 37°C) to treat or prevent CDAD is both unusual and innovative. Administering this microorganism to prevent infectious disease also avoids the deleterious effects antibiotics have upon the fecal flora.

Saccharomyces boulardii achieves high steady-state levels (10^7-10^8 colony forming units) in the colon within three days, survives gastric acid, is not absorbed, is not inhibited by antibiotics, is inhibited by standard antifungal drugs, does not significantly affect the normal flora, and is no longer detectable by 2-6 days after therapy is discontinued [1, 2, 3, 8].

Studies with *S. boulardii*

Saccharomyces boulardii is effective in treating various forms of infectious diarrhea [11], and evidence from animal (hamsters and gnotobiotic mice) models of CDAD indicates that *S. boulardii* is effective in preventing mortality due to *C. difficile* in these models [6, 7, 10, 11, 17]. An open trial enrolling patients with recurrent CDAD found that 85 percent responded to a combination treatment with vancomycin and *S. boulardii* [16].

Subsequently, three prospective, double-blind placebo-controlled studies of the efficacy of *S. boulardii* in prevention of antibiotic-associated diarrhea or recurrent *Clostridium difficile* colitis have been conducted [12, 13, 15] (Fig. 1). They concerned three significantly different patient populations. Taken together, they indicate *S. boulardii* was safe and of significant benefit in these conditions, and that it therefore shows promise of diminishing the consequences of antibiotic-associated diarrhea caused either by *C. difficile* or of unknown cause.

Prevention of Antibiotic Associated Diarrhea (AAD)

The first prospective double-blind controlled study of the efficacy of *S. boulardii* in preventing AAD was reported by Christina Surawicz and her colleagues in 1989 [15]. They evaluated the effect of live *S. boulardii* or an identical appearing placebo given in capsule form concurrently with antibiotics over a 23 month period to 180 consecutive inpatients at the Harborview Medical Center in Seattle who received new antibiotic prescriptions.

Patients were randomized in 2:1 blocks to lyophilized *S. boulardii* or placebo in 250 mg capsules taken by mouth twice daily. Studies in human volunteers indicate that this dose achieves a steady state concentration of about 10^8 viable counts per milligram of stool by the third day [13]. The regimen was initiated with 48 hours of starting antibiotic therapy and was continued for two weeks after the last antibiotic dose. Records of stool frequency and consistency were kept during hospitalization and after discharge. Stools were cultured for *C. difficile* on entry, at about day 5 and every 10 days thereafter; additional stools were cultured if the patient developed diarrhea. Stools positive for *C. difficile* were assayed for cytotoxin within 48 hours of collection. No specific attempts were made to document colitis, as the objective of the study was to evaluate the efficacy of *S. boulardii* in prevention of AAD.

Patients were excluded if they had diarrhea just before or within 24 hours of the start of the study, or immune compromised, recent chemotherapy or radiation therapy, renal failure requiring dialysis, pregnancy, treatment with antifungal drugs

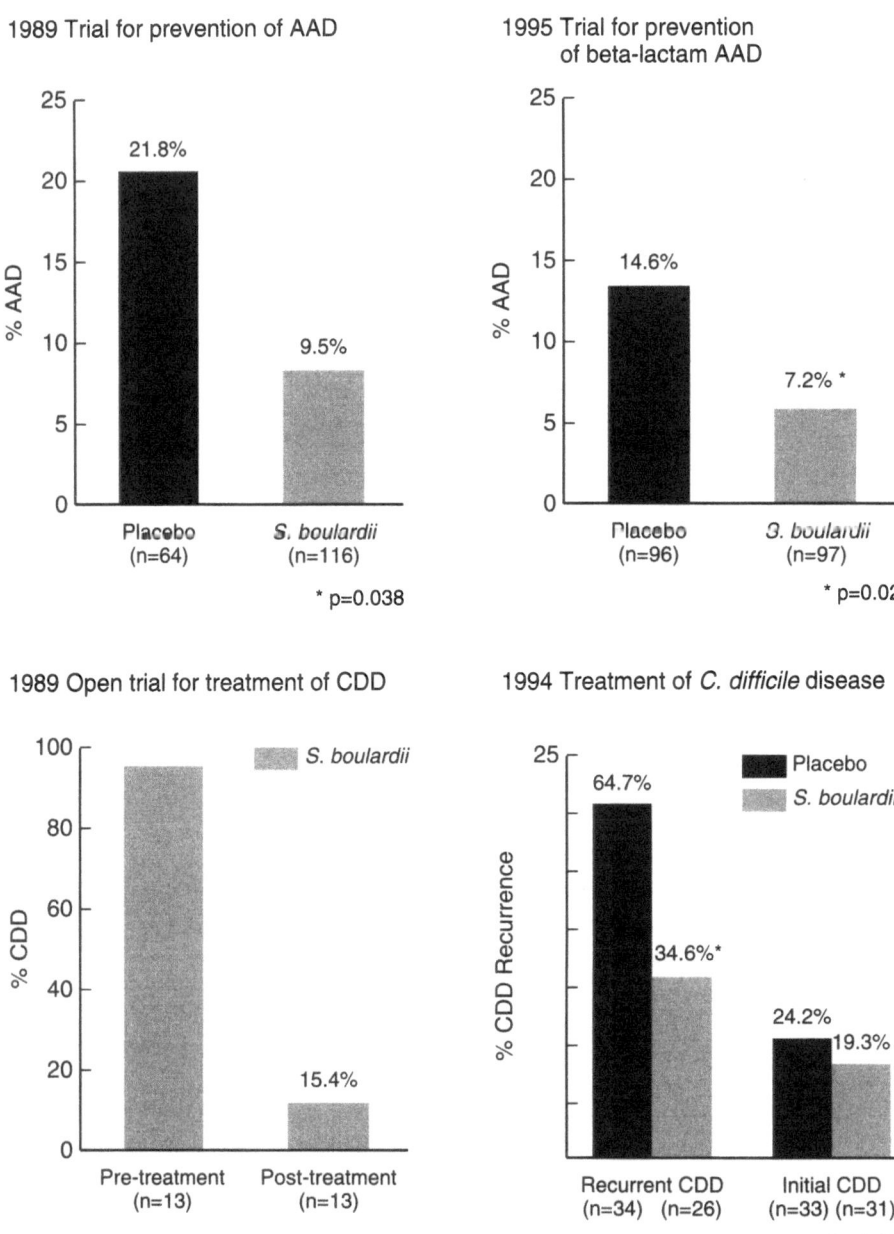

Fig. 1. A summary of four studies using the biotherapeutic agent, *Saccharomyces boulardii*. Black bars represent disease outcome (either antibiotic-associated diarrhea, AAD or *C. difficile* associated disease, CDAD) in placebo treated groups and lighter bars represent group treated with *S. boulardii*.

or lactulose, or if they were monitored for diarrhea for fewer than eight days (none of them developed diarrhea). The mean age of evaluable patients was 47.8 ± 20.1 years, and the two groups appeared comparable in all demographic factors evaluated. The results indicated there was a significant decrease in the incidence of diarrhea in patients receiving *S. boulardii*. Of the 180 patients, 14 of 64 (21.8%) on placebo developed diarrhea compared with 11 of 116 (9.5%) treated with *S. boulardii* (p = 0.038). The efficacy of *S. boulardii* in preventing AAD was 56.7 percent. In those with diarrhea, the total days of diarrhea and the percentage of patients with fever were not significantly different, but the mean number of days of diarrhea was 4.5 ± 2.9, with a range of 2-11 days. When patients on nasogastric tube feedings (a risk factor for diarrhea) were excluded from the analysis, the rate of diarrhea in the *S. boulardii* group was 5 of 109 (4.6%) compared with 13 of 59 (22%) for placebo (p = 0.001). There were no side effects of either *S. boulardii* or placebo therapy. In this study it appeared that antibiotic combinations containing clindamycin, cephalosporins, or trimethoprim-sulfamethoxazole were associated with an increased risk of diarrhea.

Of 81 patients receiving yeast, 22 (27%) acquired *C. difficile* in the hospital, while 5 (14%) of 31 receiving placebo acquired *C. difficile*. Thus, *S. boulardii* did not prevent the acquisition of *C. difficile* in these patients (p = 0.18) but it did reduce the likelihood of developing diarrhea. Of 32 *C. difficile* positive patients treated with yeast, 3 (9.4%) developed diarrhea compared with 5 (31%) of 16 receiving placebo (p = 0.07). Of the 8 patients with *C. difficile*-positive diarrhea, 3 (37.5%) were cytotoxin-positive, while of the 33 patients who were *C. difficile*-positive and who did not develop diarrhea, 16 (48.%) were cytotoxin-positive. No patients with *C. difficile*-positive diarrhea were diagnosed as having colitis.

There were many asymptomatic carriers in this study. Interestingly, of the patients without diarrhea in this study, 35% were culture-positive for *C. difficile* and nearly half of these were also positive for cytotoxin.

This study is the first report of the successful prevention of AAD using a live microorganism in an acute care setting where there was a predominance of multiple parenteral antibiotic usage. It is noteworthy that no adverse effects attributable to *S. boulardii* were discerned during this study.

The second controlled double-blind study of the efficacy of *S. boulardii* in preventing antibiotic-associated diarrhea concerned patients receiving beta-lactam antibiotics. It was reported by Lynne McFarland and her colleagues in 1995 [13], and was conducted at the University of Washington in Seattle, at the University of Kentucky in Lexington and at the St- Louis University in St. Louis, Missouri. Beta-lactam antimicrobials were chosen for study because they have been associated with an increased risk of AAD, possibly because many of them are active against the anaerobic component of the fecal flora. The general plan of the study was similar to the one reported by C. Surawicz in Seattle in 1989 and discussed previously [15]. Adult inpatients receiving new prescriptions for beta-lactam antibiotics other than penicillin G or V were eligible for the study. Patients were assigned to either oral placebo or *S. boulardii* 1 gm (3×10^{10} colony-forming units) per day (two 250 mg capsules, twice per day). Patients were randomized within three age groups at each center (18-44, 45-69, 70-99). A modified APACHE score was used to stratify acutely ill patients according to the severity of their illness. The packaging of the blinded study kits was performed in France and the investigators did not have access to the identity of the study drug, which was started within

72 hours of oral or intravenous beta-lactam antibiotics and continued for 3 days after they were discontinued. Patients were then followed for an additional seven weeks. Diarrhea was defined as a change in bowel habits with at least three loose stools/day for at least two consecutive days, and as beta-lactam associated if no other etiology of diarrhea could be identified.

Stools were cultured for *C. difficile* in the usual way, but also into prereduced supplemented peptone broth containing 0.1 percent pure sodium taurocholate so as to detect small numbers of the organism. CHO cell tissue cultures were used to detect cytotoxin B, and *C. sordellii* anti-toxin was used to check specificity. Patients were evaluated on an intention-to-treat basis.

The study enrollment period lasted from March 1989-December 1992, during which 12,546 patients were screened for entry. Of 193 eligible patients, 97 were assigned to *S. boulardii* and 96 were assigned to placebo. Patients treated with yeast received a mean of 2.4 antibiotics, while patients on placebo received a mean of 2.5. Antibiotics and other medications were not significantly different between the two groups. Of the 193 patients in the study, 21 (10.9%) experienced AAD, and seven reported diarrhea that did not meet the case definition of AAD. The mean incubation period of AAD was 18 days and ranged from 3 to 65 days (from the first day of antibiotics to the first day of diarrhea). The frequency of AAD was higher for patients on multiple antibiotics (15%) than for patients on a single beta-lactam (6%).

The incidence of AAD was significantly greater in *C. difficile*-positive patients (7/24, 29.2%) than in *C. difficile*-negative patients. Seven (7.2%) of the 97 patients receiving *S. boulardii* developed AAD compared with a significantly higher frequency (14/96, 14.6%) of patients assigned to placebo (p = 0.02). The efficacy for *S. boulardii* in preventing AAD was 51 percent. The severity of the diarrhea was not significantly different for the two groups, but the duration of diarrhea was significantly shorter for patients on *S. boulardii*. In patients with diarrhea that began after antibiotic treatment was discontinued, the duration of diarrhea fell from a median of 16 days in placebo patients to 2.5 days in *Saccharomyces boulardii* patients (p = 0.05). Five of the seven patients who received *S. boulardii* and developed diarrhea did so while receiving antibiotics, while of the 14 patients with diarrhea who received placebo, eight developed AAD while on antibiotics, and the other six had delayed diarrhea. In this study, 8/21 cases of AAD occurred during post-antibiotic exposure with incubation periods ranging from 3-46 days post-antibiotics, indicating that the human colon may require a significant time period for the normal flora to recover its protective effect. Thus, the apparent residual effects of *S. boulardii* found in this study after the yeast was discontinued may be due to the normalization of the colonic flora during *S. boulardii* administration and not to a direct action of *S. boulardii*, which is expected to be cleared from the colon a few days after it is discontinued. Of the 24 patients with positive *C. difficile* assays, the frequency of AAD was not significantly different by the study drug assignment; 3 of 10 patients on *S. boulardii* developed AAD compared with 4/14 on placebo. The Kaplan-Meier curve showed significantly more patients receiving placebo developed AAD by the end of the study compared with *Saccharomyces boulardii* (chi^2 = 3.71, p = 0.05).

Of 185 patients who returned completed adverse reaction forms, there were no significant adverse reactions except that placebo patients reported more intestinal gas and fever compared with *S. boulardii* patients. None of the patients with AAD had endoscopic examinations, so it is unknown if colitis was present. However, in patients with *C. difficile* disease, the diarrhea was sufficiently severe in 57 percent of the patients to require treatment with vancomycin or metronidazole.

Treatment of *Clostridium difficile* disease

The third multicenter prospective, randomized, placebo controlled, parallel group trial was conducted at the Universities of Washington, Kentucky and Michigan. It was reported by Lynne McFarland and colleagues in 1994 [12]. This was the first study to analyze patients with initial or recurrent *C. difficile* associated diarrheal disease (CDAD) to determine the effectiveness of a new treatment for CDAD. Its objective was to determine whether *S. boulardii* was effective in preventing recurrences of CDAD in patients experiencing either their first episode of the disease or a recurrence of the disease.

About 20 percent of patients with CDAD experience a recurrence. Even though such patients respond to treatment with vancomycin or metronidazole, once patients have had one recurrence, they may experience repeated episodes over a period of years regardless of the antibiotic used in treatment. An open trial enrolling patients with recurrent CDAD had found that 85 percent of them had no further recurrences after treatment with a combination of vancomycin and *S. boulardii* [16].

A total of 124 consenting adult patients were eligible for the study, including 64 who were having their initial episode of CDAD, and 60 who had a history of at least one prior CDAD episode. Patients who had a history of chronic diarrhea due to other bowel disease, or were pregnant, or were receiving oral antifungal therapy, or were immunosuppressed due to the acquired immunodeficiency syndrome or cancer chemotherapy within three months or were considered unreliable were excluded from the study. Each of 124 participants with active CDAD was treated with either vancomycin or metronidazole at the discretion of their primary physician, and then with lyophilized *S. boulardii* given orally as two 250 mg capsules twice per day (containing 3×10^{10} CFU/day) or placebo for a total of four weeks. Patients kept a standardized daily diary of stool frequency and consistency, other symptoms, medications and adverse reactions. At the end of study drug treatment, patients were followed for an additional four weeks. Diarrhea was defined as a change in bowel habits with at least three loose or watery bowel movements per day for at least two consecutive days. Recurrences were confirmed by positive *C. difficile* assays (culture, toxin A or toxin B). *C. difficile* colitis was defined as diarrhea with fecal white blood cells or by typical sigmoidoscopic findings. Uncomplicated diarrhea was CDAD without any evidence of inflammatory changes. Patients were evaluated on an intention to treat basis.

A total of 124 patients were eligible for the study, and 104 completed the trial. Attrition was not significantly different by study drug assignment, and the mean length of follow-up of these patients was 20.3 days. All 124 patients were included in the analysis. The study drug and placebo groups were similar with regard to baseline characteristics except that those receiving placebo had significantly more recent surgeries.

Among 33 placebo patients with their first episode of CDAD, the rate of recurrence was 24.2 percent (8/33), while placebo patients with a history of prior CDAD had a higher recurrence rate, 64.7 percent (22/34), after standard antibiotic therapy. In the 64 patients with their initial episode of CDD, treatment failure was observed in 19.3 percent of patients treated with antibiotics plus *S. boulardii* and 24.2 percent in patients treated with antibiotics plus placebo; an insignificant difference (p = 0.86). However, because of the small numbers of patients with initial CDAD who failed, there was only a 10 percent power of detecting a significant difference; therefore, there could be a type II error in this result.

The 60 patients who were entered because of a recurrence of CDAD had a mean of 3.2 prior episodes of CDAD (range, one to nine) despite having had a wide variety of prior treatments. In 60 patients with recurrent CDAD, the rate of recurrence with *S. boulardii* was significantly lower, 34.6 percent (9/26), than it was among placebo recipients, 64.7 percent (22/34) (p = 0.04), resulting in an efficacy of 46.5 percent. Using a logistic regression model to control for significant confounding factors, the relative risk of treatment failure with *S. boulardii* compared with placebo was 0.43 (95% confidence interval, 0.20 to 0.97).

Overall, of the 124 total patients enrolled in the study, 30 (44.8 %) of 67 patients receiving placebo failed treatment compared with 15 (26.3 %) of 57 patients receiving *S. boulardii* (chi^2 = 3.78, p = 0.05). The efficacy of *S. boulardii* in combination with standard oral antibiotic treatment for the prevention of CDAD recurrence in the 124 patients was 41.3 percent. The effectiveness of *S. boulardii* was similar in patients with uncomplicated diarrhea (47.3%), colitis (37.5%), or pseudomembranous colitis (46.6%).

Using multivariate analyses, dosages or treatment with either vancomycin or metronidazole in the two study groups were not associated with an increased risk of recurrence. Of 65 patients receiving only vancomycin, 28 (43 %) failed, and the mean dosage was not significantly different in those who failed and those who did not. Of the 37 receiving only metronidazole, 12 (32%) failed, and the mean dosage was not different in those who failed and those who did not. Neither the severity of the enrollment CDAD episode nor the type of standard antibiotic therapy given was found to be a significant risk factor for recurrence. Using a multiviariate model, only a history of prior CDD was found to be a significant factor for CDAD recurrence. The severity of the recurrence was analyzed in the 45 patients who failed combination therapy. Patients receiving *S. boulardii* had significantly fewer daily stools (mean of 2.1) than patients receiving placebo (mean of 3.3, p = 0.02), but there was no difference in abdominal pain, cramps or nausea in the two study groups. The time of recurrence was similar in the two study drug groups in that most occurred 1 to 2 weeks after antibiotic therapy was discontinued. Of the 15 CDAD recurrences in patients treated with *S. boulardii*, 13 (87 %) failed during the study drug period (weeks 1-4) and two (13%) failed during the surveillance period. Of the 30 recurrences in patients treated with placebo, 21 (70 %) failed during the study drug period and nine (30 percent) failed during surveillance.

At the end of standard antibiotic therapy, 10 (19%) of the patients receiving placebo and 6 (12%) of those receiving *S. boulardii* were still positive for *C. difficile*; its persistence was not predictive of an increased risk of subsequent recurrence. At the end of the four weeks of study drug treatment, only three (8.6 %) of 35 tested patients who received *S. boulardii* continued to harbor *C. difficile* (as measured by culture or toxin test) compared with 11 (26.8 %) of those receiving placebo (p = 0.04). *S. boulardii* did not significantly reduce culture positivity, but did significantly reduce the frequency of toxin B positivity (6.7 %) compared with placebo (30 %) (p = 0.02) by the end of week 4.

Two adverse reactions were reported more frequently during the study drug period by patients receiving *Saccharomyces boulardii* than with placebo: five patients noted an increase in thirst while receiving *S. boulardii* while none on placebo reported this symptom (p = 0.02), and eight reported constipation on *S. boulardii* compared with two on placebo. No adverse reactions to *S. boulardii* were observed during follow-up.

Conclusions

The important conclusions of this study were that the combination of standard antibiotics and *Saccharomyces boulardii* was shown to be a safe and effective therapy for patients with recurrent *Clostridium difficile* diarrheal disease, but it was not significantly effective for patients with their first episode of the disease, possibly because both the rate of recurrence was low in the placebo group and the number of patients studied was small. The mechanism of action of *S. boulardii* in this study is not fully elucidated, but a recent study reported that it produces a protease that interferes with the binding of toxin A to specific receptors in rat ileum [14]. Another recent study found that *S. boulardii* may prevent diarrhea by stimulating chloride absorption [9]. In addition, *S. boulardii* exerts trophic effects on the intestinal mucosa, resulting in an increase in secretory component, secretory IgA [16], and in intestinal enzymes such as lactase, maltase and sucrase [17]. In the present study, *S. boulardii* was found to significantly decrease the frequency of toxin B positivity by week 4 of therapy with it, but not the rate of *C. difficile* culture positivity, which supports an earlier study with *S. boulardii* in hamsters with colitis that showed significant decreases in toxin A and B positivity, but not in stool concentrations of *C. difficile* [7].

References

1. Berg R, Bernasconi P, Fowler D, Gautreau M (1993) Inhibition of *Candida albicans* translocation from the gastrointestinal tract of immunosuppressed mice by oral treatment with *Saccharomyces boulardii*. J Infect Dis 168: 1314-1318
2. Blehaut H, Massot J, Elmer GW, Levy RH (1989) Disposition kinetics of *Saccharomyces boulardii* in man and rat. Biopharm Drug Dispos 10: 353-364
3. Boddy AV, Elmer GW, McFarland LV, Levy RH (1991) Influence of antibiotics on the recovery and kinetics of *Saccharomyces boulardii* in rats. Pharm Res 8: 796-800
4. Buts JP, Bernasconi P, Vaerman JP, Divers C (1990) Stimulation of secretory IgA and secretory component of immunoglobulins in small intestine of rats treated with *Saccharomyces boulardii*. Dig Dis Sci 35: 251-256
5. Buts JP, Bernasconi P, Van Craynest MP, Maldague P, De Meyer R (1986) Response of human and rat small intestinal mucosa to oral administration of *Saccharomyces boulardii*. Pediatr Res 20: 192-196

6. Corthier G, Dubos, Ducluzeau R (1986) Prevention of *C. difficile* induced mortality in gnotobiotic mice by *Saccharomyces boulardii*. Can J Microbiol 32: 894-896

7. Elmer GW, McFarland LV (1987) *Saccharomyces boulardii* suppression of overgrowth of toxigenic *Clostridium difficile* after vancomycin treatment in hamsters. Antimicrob Agents Chemother 31: 129-131

8. Klein SM, Elmer GW, McFarland LV, et al (1993) Recovery and elimination of the biotherapeutic agent, *Saccharomyces boulardii*, in healthy human volunteers. Pharm Res 10: 1615-1619

9. Krammer M, Karbach U (1993) Antidiarrheal action of the yeast *Saccharomyces boulardii* in the rat small and large intestine by stimulating chloride absorption. Z Gastroenterol 31(suppl 4): 73-77

10. Massot J, Sanchez O, Couchy R, et al (1984) Bacteriopharmacological activity of *Saccharomyces boulardii* in clindamycin induced colitis in the hamster. Arnzeimittelforschung 34: 794-797

11. McFarland LV, Bernasconi P (1993) A review of a novel biotherapeutic agent: *Saccharomyces boulardii*. Microb Ecol Health Dis 6: 157-171

12. McFarland LV, Surawicz CM, Greenberg RN, Fekety R, Elmer GW, Moyer K, Melcher SA, Bowen KE, Cox J, Noorani Z, Hamilton G, Rubin M, Greenwald D (1994) A randomized placebo-controlled trial of *Saccharomyces boulardii* in combination with standard antibiotics for *Clostridium difficile* disease. JAMA 271: 1913-1918

13. McFarland, LV, Surawicz CM, Greenberg, RN, et al (1995) Prevention of ß-Lactamassociated diarrhea by *Saccharomyces boulardii* compared with placebo. Am J Gastroenterol 90: 439-448

14. Pothoulakis C, Kelly CP, Joshi MA, Gao N, O'Keane CJ, Castagliuolo I, LaMont JT (1993) *Saccharomyces boulardii* inhibits *Clostridium difficile* toxin A binding and enterotoxicity in rat ileum. Gastroenterology 104: 1108-1115

15. Surawicz CM, Elmer GW, Speelman P, McFarland LV, Chinn J, van Belle G (1989) Prevention of antibiotic-associated diarrhea by *Saccharomyces boulardii*: A prospective study. Gastroenterology 96: 981-988

16. Surawicz CM, McFarland LV, Elmer G, Chinn J (1989) Treatment of recurrent *Clostridium difficile* colitis with vancomycin and *Saccharomyces boulardii*. Am J Gastroenterol 84: 1285-1287

17. Toothaker RD, Elmer GW (1984) Prevention of clindamycin-induced mortality in hamsters by *Saccharomyces boulardii*. Antimicrob Agents Chemother 26: 552-556

Mechanisms of *Saccharomyces boulardii* on *Clostridium difficile* infection

Charalabos Pothoulakis

SUMMARY

Saccharomyces boulardii is a non-pathogenic yeast effectively used in the prevention and treatment of *Clostridium difficile*-mediated diarrhea and colitis. Several putative mechanisms for the protective effects of *S. boulardii* have been proposed based on results from human as well as animal studies for *C. difficile* diarrhea and intestinal inflammation. Treatment of humans and rats with *S. boulardii* increases the intestinal levels of the disaccharidases maltase and lactase. Subsequent studies indicated that polyamines released from *S. boulardii* itself may cause these trophic effects on the small intestine. Administration of *S. boulardii* to rats also increases the intestinal levels of the secretory component of immunoglobulins and of secretory IgA, indicating that this yeast may exert an immunoprotective effect against *C. difficile* and its toxins. Pretreatment of animals with *S. boulardii* protects from death following *C. difficile* infection and this effect could be mediated by a yeast effect on *C. difficile* bacterium as well as its toxins A and B. Results from our laboratory showed that treatment of rats with *S. boulardii* before addition of toxin A to rat intestine substantially reduced the enterotoxic effects of toxin A. Further experiments indicated that these effects of *S. boulardii* are mediated, at least in part, by a yeast secreted protease which is able to digest toxin A as well as its intestinal receptor *in vitro*. These results suggest that *S. boulardii* reduces the intestinal effects of *C. difficile* toxin A by digesting the toxin A molecule as well as its intestinal brush border receptor.

Taken together, results from our as well as other laboratories provide several possible mechanisms for the action of *S. boulardii* on human *C. difficile* infection. These include: (1) proteolysis of toxin A and its intestinal receptor by a yeast protease, (2) a direct inhibition of *C. difficile* growth, (3) enhancement of an intestinal immune response and (4) stimulation of intestinal disaccharidase activity possibly through release of polyamines by *S. boulardii*.

Mécanismes d'action de *Saccharomyces boulardii* au cours des infections à *Clostridium difficile*

RÉSUMÉ

Saccharomyces boulardii est une levure non pathogène utilisée dans la prévention et le traitement des diarrhées et colites dues à *Clostridium difficile*. Plusieurs mécanismes pouvant expiquer les effets protecteurs de *S. boulardii*, ont été proposés à partir de l'étude chez l'homme ou l'animal des diarrhées ou infections intestinales associées à *C. difficile*. Le traitement par *S. boulardii* d'humains ou de rats augmente le taux intestinal de disaccharidases telles que la maltase ou la lactase. Des études complémentaires ont montré que la libération de polyamines par *S. boulardii*, pouvait expliquer cet effet trophique sur l'intestin grêle. L'administration de *S. boulardii* à des rats augmente également le taux intestinal de la pièce sécrétoire des immunoglobulines et des IgA sécrétoires, indiquant que cette levure exerce un effet immunoprotecteur contre *C. difficile* et ses toxines. Le prétraitement d'animaux par *S. boulardii* les protège de la mort consécutive à une infection par *C. difficile*, et cet effet peut être médié par un effet de la levure sur la bactérie de *C. difficile* comme sur ses toxines A et B. Nos résultats montrent que si l'on traite des rats avec *S. boulardii* avant de leur administrer la toxine A au niveau intestinal, les effets entérotoxiques de la toxine A sont sensiblement réduits. Des expériences ultérieures précisent que ces effets de *S. boulardii* sont médiés, au moins en partie, par une protéase, sécrétée par la levure, et capable de digérer la toxine A aussi bien que son récepteur intestinal *in vitro*. Ces résultats suggèrent que *S. boulardii* réduit les effets intestinaux de la toxine A de *C. difficile* en digérant la molécule de toxine A, mais aussi son récepteur au niveau de la bordure en brosse intestinale.

L'ensemble des données émanant de notre laboratoire, comme celles d'autres équipes, fournissent plusieurs hypothèses concernant le mécanisme d'action de *S. boulardii* lors d'une infection à *C. difficile* chez l'homme. Ces hypothèses sont les suivantes : (1) protéolyse de la toxine A et de son récepteur intestinal par une protéase sécrétée par la levure; (2) inhibition directe de la croissance de *C. difficile*; (3) stimulation de la réponse immunitaire intestinale; (4) stimulation de l'activité des disaccharidases intestinales, probablement en raison d'une libération de polyamines par *S. boulardii*.

Wirkmechanismen von *Saccharomyces boulardii* bei intestinalen *Clostridium difficile*-Infektionen

ZUSAMMENFASSUNG

Saccharomyces boulardii ist eine apathogene Hefe, die erfolgreich zur Prävention und Behandlung von *C. difficile*-vermittelter Diarrhö und Kolitis eingesetzt wird. Ausgehend von den Ergebnissen am Menschen und im Tierversuch wurden mehrere mutmaßliche Mechanismen für die Schutzwirkung von *S. boulardii* gegenüber *C. difficile*-Diarrhöen und Darmentzündungen vorgestellt. Bei Behandlung mit *S. boulardii* steigen bei Menschen und Ratten die Spiegel der Disaccharidasen Maltase und Lactase im Darm an. Anschließende Studien zeigten, daß von *S. boulardii* selbst produzierte Polyamine möglicherweise für die trophische Wirkung im Dünndarm verantwortlich ist. Die Gabe von *S. boulardii* erhöhte bei Ratten darüber hinaus die Spiegel der sekretorischen Komponente von Immunglobulinen und sekretorischem IgA im Darm, was eine immunprotektive Wirkung der Hefe gegenüber *C. difficile* und seinen Toxinen vermuten läßt. Im Tierversuch schützte eine Vorbehandlung mit *S. boulardii* nach einer Infektion mit *C. difficile* die Tiere vor dem Tod; dieser Effekt könnte auf eine Wirkung der Hefe auf das *C. difficile*-Bakterium oder auf seine Toxine A und B zurückzuführen sein. Die Ergebnisse unseres Labors zeigten, daß die Behandlung von Ratten mit *S. boulardii* vor dem Einbringen von Toxin A in den Rattendarm die enterotoxischen Wirkungen dieses Toxins erheblich verringerte. Weitere Experimente zeigten, daß diese Wirkungen von *S. boulardii* zumindest teilweise durch eine von der Hefe sezernierte Protease vermittelt werden, die *in vitro* in der Lage ist, Toxin A und den entsprechenden Darmrezeptor zu verdauen. Diese Befunde legen die Vermutung nahe, daß *S. boulardii* die intestinalen Wirkungen des *C. difficile*-Toxins A dadurch abmildert, daß es das Toxin A-Molekül ebenso wie seinen Rezeptor im intestinalen Bürstensaum verdaut. Insgesamt zeigen unsere Ergebnisse ebenso wie die anderer Labors mehrere mögliche Mechanismen für die Wirkung von *S. boulardii* gegen *C. difficile*-Infektionen beim Menschen auf, nämlich (1) Proteolyse von Toxin A und seinem

Darmrezeptor durch eine von der Hefe sezernierte Protease, (2) eine direkte Hemmung der *C. difficile*-Vermehrung, (3) eine Verstärkung der intestinalen Immunreaktion und (4) eine Stimulierung der Aktivität intestinaler Disaccharidasen, möglicherweise durch die Freisetzung von Polyaminen durch *S. boulardii*.

Introduction

Clostridium difficile mediated antibiotic-associated colitis and diarrhea is one of the most frequent nosocomial infections in the US and Europe [19]. The pathophysiology of this infection involves alterations of the colonic flora by antibiotics, ingestion of spores, and colonization of the large intestine by *C. difficile*, followed by the release of its toxins, toxin A and toxin B [15]. It is well established that *C. difficile* infection is a toxin-mediated disease. Animal studies indicated that toxin A, but not toxin B, causes fluid secretion, mucosal damage and release of inflammatory mediators when injected into rodent intestine [22, 27]. In humans, however, both toxins cause mucosal damage and electrophysiologic changes when applied to the colonic mucosa. Toxin B is more potent than toxin A in inducing these responses [24], indicating that both toxins mediate *C. difficile* colitis in humans.

Therapy for *C. difficile* can be aimed at any point in the above described pathophysiologic sequences. A unique therapeutic approach is probiotic therapy, defined as the use of a non pathogenic organism to compete with a pathogenic one. *S. boulardii* is a non-pathogenic yeast used empirically in Europe for treatment of infectious diarrhea. One of the major applications of *S. boulardii* has been the prevention and treatment of antibiotic-associated diarrheal diseases, including *C. difficile* associated diarrhea and colitis. Two double-blind controlled prospective studies indicated that the frequency of antibiotic - related diarrhea was significantly less frequent in patients receiving *S. boulardii* compared with patients treated with a placebo [20, 25]. McFarland et al [21] recently reported that only 24% of patients with *C. difficile* disease receiving *S. boulardii* in combination with vancomycin or metronidazole experienced recurrence as compared to 45% of patients who received these antibiotics without *S. boulardii*. In a recent uncontrolled study Buts et al [3] reported a significant symptomatic improvement in 16 of 19 children treated with *S. boulardii* for acute *C. difficile* infection. In addition, treatment with *S. boulardii* resulted in clearing of stool cytotoxin of *C. difficile* in 85% of these patients [3]. Possible mechanisms that could account for the *S. boulardii* protective effects have been suggested by several laboratories based on results from animal models of *C. difficile*, including (1) trophic effects on the intestinal mucosa, (2) enhancement of the host intestinal immune response, (3) effects on the *C. difficile* bacterium and/or its toxins, and (4) effects on the toxin receptors. This paper will discuss these putative mechanisms.

Effects on the intestinal mucosa

Treatment of humans with *S. boulardii* for 14 days resulted in a marked increase of the duodenal and jejunal levels of the intestinal disaccharidases sucrase, maltase and lactase as compared to the levels of the enzymes before treatment [2] (Fig. 1). Similar results were obtained in the jejunum of rats treated with either viable or heat-killed *S. boulardii* for 5 days as compared to control rats which received only saline [2]. In a subsequent study, Buts et al [4] studied the possibility that poly-amines, bacterial and yeast products known to modulate cellular division and protein synthesis, are responsible for the enhancement of intestinal disaccharidase activity caused by *S. boulardii* [2]. Administration of *S. boulardii* in weaning rats resulted in significant increases in the intestinal levels of the polyamines spermine and spermidine [4]. Since lyophilized preparations of *S. boulardii* produced substantial amounts of these polyamines and administration of spermine and spermidine to rats stimulated intestinal disaccharidase activity [4], these results indicate that *S. boulardii* may cause its trophic effects on the small intestine by the release of polyamines.

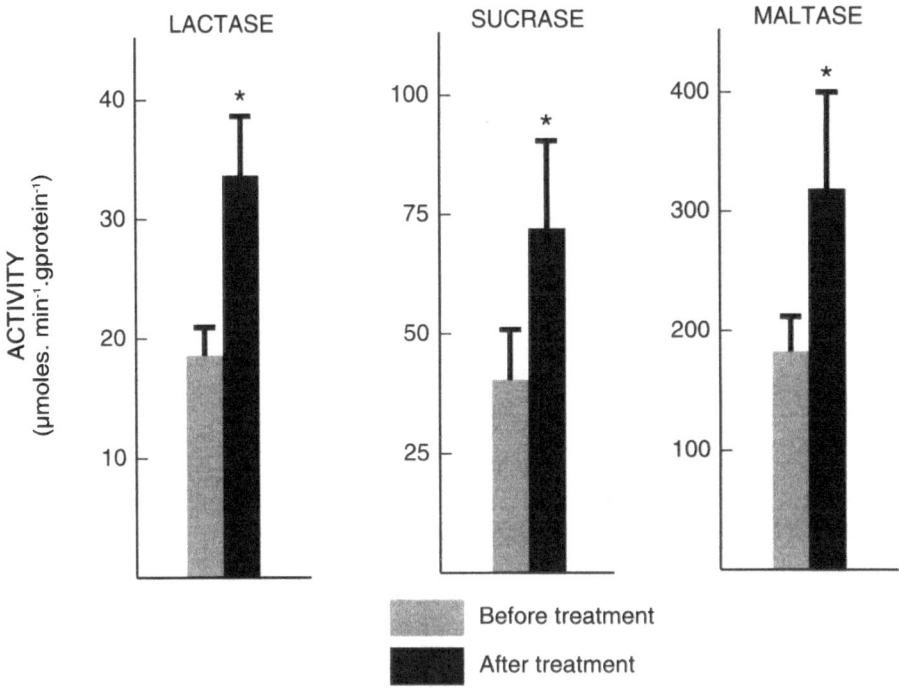

Fig. 1. Changes in mucosal specific activity of sucrase, lactase, and maltase after 14 days of oral treatment with *S. boulardii* in seven volunteers. *, p < 0.05 [from Buts et al (1986) Pediatr Res 20: 192-196, with permission]

Immunoprotective effects

Administration of *S. boulardii* to weaning rats significantly increased the duodenal levels of the secretory component of immunoglobulins (by 62.8%) and of secretory IgA (by 56.9%) as compared to levels of these immunoglobulins in control rats [1] (Fig. 2). Furthermore, these changes did not appear to be related to an increased turnover rate of intestinal cells in response to the yeast since DNA enterocyte synthesis was similar between *S. boulardii* treated and control animals [1]. These results indicate that *S. boulardii* may exert an immunoprotective effect against bacterial pathogens, including *C. difficile* and its toxins by stimulating production of secretory IgA. In addition to its effects on immunoglobulin production, *in vitro* studies showed that *S. boulardii* can directly activate complement and fix C3b [16]. Furthermore, phagocytosis of *S. boulardii* by human mononuclear cells appear to be complement-dependent [16].

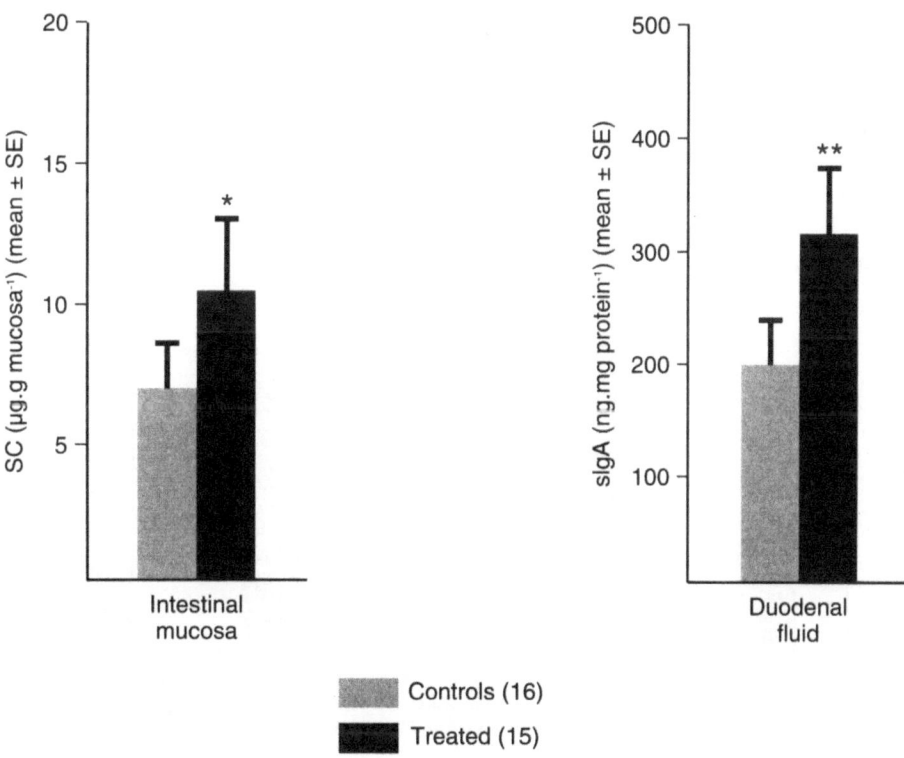

Fig. 2. Increased levels of the secretory component of immunoglobulins (SC, left) and secretory IgA (s IgA, right) in the intestinal mucosa and the duodenal fluid of *S. boulardii*-treated rats compared to saline controls. *p < 0.05, and **p < 0.01 [from Buts et al (1990) Dig Dis Sci 35: 251-256, with permission]

Effects on *Clostridium difficile* and its toxins

Pretreatment of animals with *S. boulardii* protected from toxigenic *C. difficile*. For example, Corthier et al [8] showed that a single *S. boulardii* ingestion protected 16% of mice, from death following challenge with *C. difficile*, whereas 54% of mice were protected when *S. boulardii* was given continuously into their drinking water. A later study indicated that this protective effect was dependent on the dose of *S. boulardii* [13] and similar protective effect was also noted in Syrian hamsters treated with *S. boulardii* before clindamycin challenge [18, 26].

    S. boulardii may protect from *C. difficile* in these animal models by inhibiting either growth of *C. difficile* or production of toxins by the bacteria or both. Although several studies showed that *S. boulardii* treatment decreases the intestinal levels of *C. difficile* bacteria in hamsters [14,18], other studies failed to confirm an inhibitory effect [6].

    S. boulardii appears to affect the levels of toxins A and B produced by *C. difficile*. Successful treatment of human [3] and experimental [6,7] *C. difficile* infection with *S. boulardii* resulted in significantly decreased stool toxin levels and this effect was accompanied by a reduction of the mucosal cell damage and intestinal inflammation associated with these toxins. A similar protective effect for

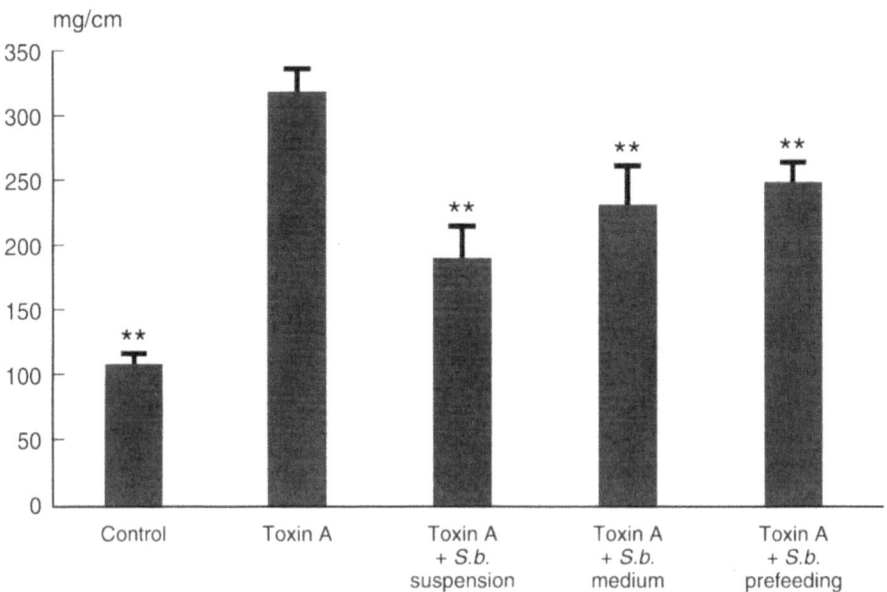

Fig. 3. *S. boulardii* reduces toxin A - induced fluid secretion in rat ileum. Ligated ileal loops were injected with buffer containing either toxin A or buffer alone or *S. boulardii* suspension or *S. boulardii* conditioned medium mixed with toxin A. In another set of experiments, rats were provided with *S. boulardii* suspension for 3 days, ligated loops were then formed and injected with toxin A. Ileal secretion was measured after 4 hours. *, p < 0.05, and **, p < 0.01 vs toxin A [from Pothoulakis et al (1993) Gastroenterology 104: 1108-1115, with permission]

S. boulardii on *C. difficile* toxin A mediated enterotoxicity *in vivo* was noted in studies from our laboratory. For example, treatment of rats with *S. boulardii* suspension for 3 days before addition of purified toxin A to ileal loops significantly reduced intestinal secretion and ^{3}H-mannitol permeability caused by this toxin [23] (Fig. 3). We also showed a substantial reduction of fluid secretion and intestinal permeability when *S. boulardii* suspension or *S. boulardii* conditioned medium was coadministered with toxin A [23], indicating that a secreted yeast factor may be responsible for these effects. Preincubation of toxin A with *S. boulardii* suspension did not alter the morphological effects of the toxin on rounding of cells in culture [23], in keeping with similar results reported by Corthier et al [9]. Interestingly, *S. boulardii* may have a similar effect on other toxin-mediated diarrheas. For example, addition of *S. boulardii* to rat jejunal loops preincubated with cholera toxin significantly inhibited water and sodium secretion caused by the toxin [28]. Recent results indicated that a 120 kD *S. boulardii* protein may be responsible for the inhibitory effect of the yeast on cholera toxin-mediated secretion [11]. In another study, Massot et al [17] showed that coadministration of *S. boulardii* suspension with a toxin-producing *E. coli* strain into mouse intestinal loops also inhibited diarrhea caused by this toxin.

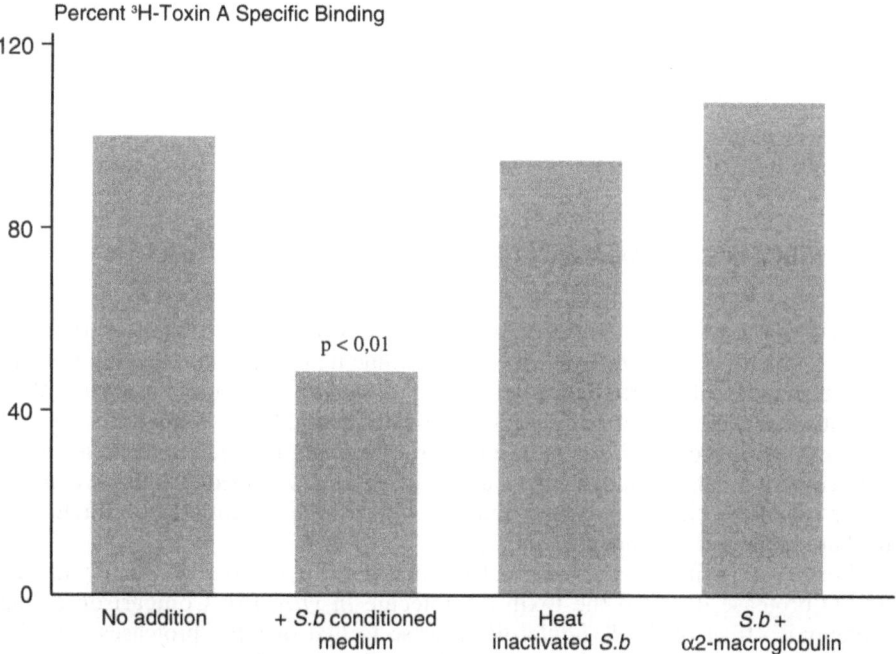

Fig. 4. *S. boulardii* treatment reduces ^{3}H-toxin A specific binding to intestinal brush border membranes. This inhibitory effect is abolished by heat-inactivation of *S. boulardii* conditioned medium or by preincubation of *S. boulardii* conditioned medium with the protease inhibitor α2-macroglobulin [from Pothoulakis et al (1993) Gastroenterology 104: 1108-1115, with permission]

Effect on toxin A receptor

Binding of the toxin to its brush border receptor is required for expression of enterotoxicity [10, 12, 29]. We therefore studied the possible effect of *S. boulardii* on toxin A receptor binding. Our results showed that preincubation of rat intestinal brush borders with *S. boulardii* conditioned medium reduced radiolabelled toxin A receptor binding in a dose dependent fashion [23]. Under the same conditions, heat inactivated *S. boulardii* conditioned medium or *S. boulardii* conditioned medium preincubated with the protease inhibitor α2 macroglobulin had no significant inhibitory effect as shown in Fig. 4 [23].

In addition, preincubation of conditioned medium from *S. boulardii* with partially purified toxin A receptor also inhibited binding of ³H-toxin A to this receptor. Since previous results indicated that the toxin A intestinal receptor is a trypsin-sensitive glycoprotein, we examined whether *S. boulardii* enzymatically altered the toxin A receptor. Polyacrylamide gel electrophoresis of purified intestinal brush borders exposed to *S. boulardii* revealed progressive diminution of all brush border proteins and disappearance of others as compared to brush borders exposed to heat inactivated *S. boulardii* conditioned medium [23]. Preincubation of *S. boulardii* with the protease inhibitor α2 macroglobulin abolished this protease-like effect of the yeast on brush border proteins [23]. Taken together, these results suggest that *S. boulardii* reduces the intestinal effects of *C. difficile* toxin A by inhibiting binding of the toxin to its intestinal brush border receptor. Furthermore, our results indicate that the effects of *S. boulardii* on toxin A receptor binding may be mediated by a heat sensitive factor secreted by the yeast which possesses protease activity.

Purification of the *S. boulardii* protease

Subsequent studies were undertaken in our laboratory in order to purify and characterize the *S. boulardii* protease and examine its mechanism of action in toxin A-mediated enteritis in rat ileal loops *in vivo*. *S. boulardii* protease was purified by gel filtration using a combination of G-50 Sephadex and Octyl-Sepharose columns. A 54 kD serine protease inhibited the toxin's effects on fluid secretion and mannitol permeability in rat ileal loops *in vivo* and reduced toxin A - mediated tissue damage of the rat villus epithelium, congestion and edema of the mucosa and infiltration of the lamina propria with neutrophils [5].

Further experiments showed that pretreatment of purified toxin A with the 54 kD protease digested the toxin A molecule *in vitro* [5]. Comparison of the proteolytic effect of the *S. boulardii* protease to that of other proteases (trypsin, chymotrypsin, carboxylpeptidase B) indicated that not all proteases were able to digest the toxin and that *S. boulardii* protease, among those tested, was the most efficient in hydrolyzing toxin A [5]. The purified protease also reduced binding of radiolabelled toxin A to its brush border receptor *in vitro*, thus confirming our previous results [23]. These results indicate that this protease may inhibit the intestinal effects of toxin A by exerting an enzymatic effect on both the toxin A molecule and its intestinal brush border receptor.

In addition, these results suggest that the protective effect of *S. boulardii* conditioned medium in the rat ileal loop model is attributable largely or entirely to the 54 kD yeast serine protease that hydrolyzes toxin A as well as its brush border glycoprotein receptor. Since toxin A from *C. difficile* is released into the bowel lumen during clinical infection [15] and since the *S. boulardii* protease would also be released into the gut lumen following oral ingestion, we propose that gradual proteolysis by *S. boulardii* protease of toxin A (and probably toxin B) and of their colonic receptors would occur, thus reducing the effects of these potent toxins on the mucosa (Table 1).

Table 1. Mechanism of *S. boulardii* on experimental *C. difficile*-mediated enteritis

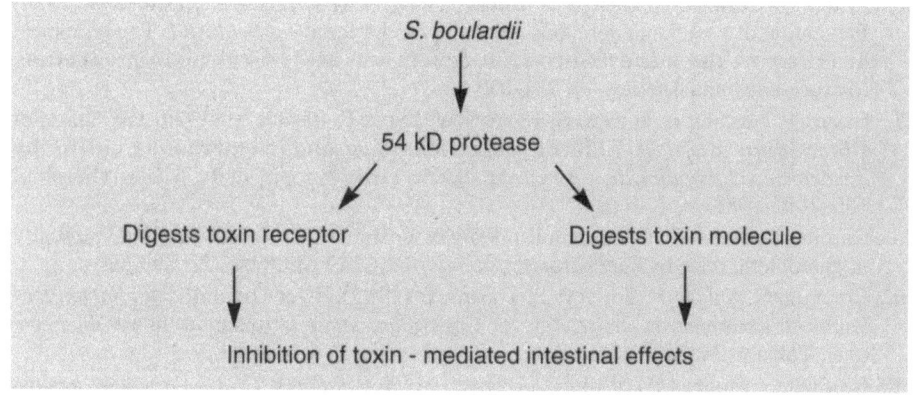

In summary, results from studies discussed in this paper provide several plausible mechanisms for the therapeutic action of *Saccharomyces boulardii* in patients infected with *Clostridium difficile*. Results from our laboratory indicate that the yeast releases a 54 kD serine protease that is capable of hydrolyzing toxin A and its intestinal receptor thus diminishing the diarrhea and other intestinal effects of the toxin. In the experimental model used in our studies, the yeast protease appears to account for most of the protective action present in conditioned yeast medium. Other possible actions of *S. boulardii*, also include direct inhibition of *C. difficile* growth, stimulation of a secretory immune response, and enhancement of intestinal disaccharidase activity through release of polyamines by *S. boulardii*.

References

1. Buts JP, Bernasconi P, Vaerman JP, Divers C (1990) Stimulation of secretory IgA and secretory component of immunoglobulins in small intestine of rats treated with *Saccharomyces boulardii*. Dig Dis Sci 35: 251-256

2. Buts JP, Bernasconi P, Van Craynest M, Maldague P, DeMeyer R (1986) Response of human and rat intestinal mucosa to oral administration of *Saccharomyces boulardii*. Pediatr Res 20: 192-196

3. Buts JP, Corthier G, Delmée M (1993) *Saccharomyces boulardii* for *Clostridium difficile*-associated enteropathies in infants. J Pediatr Gastroenterol Nutr 16: 419-425

4. Buts JP, Keyser ND, Raedemaeker LD (1994) *Saccharomyces boulardii* enhances rat intestinal enzyme expression by endoluminal release of polyamines. Pediatr Res 36:522-527

5. Castagliuolo I, LaMont JT, Baker C, Guzman DD, Nikulasson ST, Pothoulakis C (1995) Purified *Saccharomyces boulardii* protease inhibits *C. difficile* toxin A effects in rat ileum (Abstract). Gastroenterology 108: A792

6. Castex F, Corthier G, Jouvert S, Elmer GW, Guibal J, Lucas F, Bastidel M (1989) Prevention of pseudomembranous cecitis by *Saccharomyces boulardii*: Topographical histology of the mucosa, bacterial counts and analysis of toxin production. Microecology and Therapy 19: 241-250

7. Castex F, Corthier G, Jouvert S, Elmer GW, Lucas F, Bastide M (1990) Prevention of *Clostridium difficile*-induced experimental pseudomembranous colitis by *Saccharomyces boulardii*: a scanning electron microscopic study. J Gen Microbiol 136: 1085-1089

8. Corthier G, Dubos F, Ducluzeau R (1986) Prevention of *Clostridium difficile* mortality in gnotobiotic mice by *Saccharomyces boulardii*. Can J Microbiol 32: 894-896

9. Corthier G, Lucas F, Jouvert S, Castex F (1992) Effect of oral *Saccharomyces boulardii* treatment on the activity of *Clostridium difficile* toxins in mouse digestive tract. Toxicon 30: 1583-1589

10. Corthier G, Muller MC, Wilkins TD, Lyerly D, L'Haridon R (1991) Protection against experimental pseudomembranous colitis in gnotobiotic mice by use of monoclonal antibodies against *Clostridium difficile* toxin A. Infect Immun 59: 1192-1195

11. Czerucka D, Roux I, Rampal P (1994) *Saccharomyces boulardii* inhibits secretagogue-mediated cAMP induction in cultured intestinal cells. Gastroenterology 106: 65-72

12. Eglow R, Pothoulakis C, Itzkowitz S, Israel EJ, O'Keane CJ, Gong D, Gao N, Xu L, Walker A, LaMont JT (1992) Diminished *Clostridium difficile* toxin A sensitivity in newborn rabbit ileum is associated with decreased toxin A receptor. J Clin Invest 90: 822-829

13. Elmer GW, Corthier G (1990) Modulation of *Clostridium difficile* induced mortality as a function of the dose and the viability of the *Saccharomyces boulardii* used as a preventive agent in gnotobiotic mice. Can J Microbiol 37: 315-317

14. Elmer GW, McFarland LV (1987) Suppression by *Saccharomyces boulardii* of toxigenic *Clostridium difficile* overgrowth after vancomycin treatment in hamsters. Antimicrob Agents Chemother 31: 129-131

15. Kelly CP, Pothoulakis C, LaMont JT (1994) *Clostridium difficile* colitis. N Engl J Med 330: 257-262

16. Machado Catano JA, Parames MT, Babo MJ, Santos A, Bandeira Ferreira A, Freitas AA, Clementa Coelho MR, Matthioli Mateus A (1986) Immunopharmacological effects of *Saccharomyces boulardii* in healthy volunteers. Int J Immunopharmacol 8: 245-259

17. Massot J, Descoinclois M, Astoin J (1982) Protection par *Saccharomyces boulardii* de la diarrhée à *Escherichia coli* du souriceau. Ann Pharm Fr 40:445-449

18. Massot J, Sanchez O, Astoin R, Parodi AL (1984) Bacterio-pharmacological activity of *Saccharomyces boulardii* in Clindamycin-induced colitis in the hamster. Arzneim-Forch/Drug Res 34: 794-797

19. McFarland LV, Mulligan ME, Kwok RYY, Stamm WE (1988) Nosocomial acquisition of *Clostridium difficile* infection. N Engl J Med 320: 204-210

20. McFarland LV, Surawicz CM, Elmer GW, Moyer KA, Melcher SA, Greenberg R, Bowen K (1993) Multivariate analysis of the clinical efficacy of a biotherapeutic agent, *Saccharomyces boulardii* for the prevention of antibiotic-associated diarrhea (Abstract). Am J Epidemiol 138: 649

21. McFarland LV, Surawicz CM, Greenberg RN, Fekety R, Elmer GW, Moyer K, Melcher SA, Bowen KE, Cox J, Noorani Z, Hamilton G, Rubin M, Greenwald D (1994) A randomized placebo-controlled trial of *Saccharomyces boulardii* in combination with standard antibiotics for *Clostridium difficile* disease. JAMA 271: 1913-1918

22. Mitchell TJ, Ketley JM, Haslam SC, Stephen J, Burdon DW, Candy DCA, Daniel R (1986) Effect of toxin A and B of *Clostridium difficile* on rabbit ileum and colon. Gut 27:78-85

23. Pothoulakis C, Kelly CP, Joshi MA, Gao N, O'Keane CJ, Castagliuolo I, LaMont JT (1993) *Saccharomyces boulardii* inhibits *Clostridium difficile* toxin A binding and enterotoxicity in rat ileum. Gastroenterology 104: 1108-1115

24. Riegler M, Sedivy R, Pothoulakis C, Hamilton G, Zacheri J, Biscof G, Consetini E, Feil W, Schiessel R, LaMont JT, Wenzl E (1995) *Clostridium difficile* toxin B is more potent than toxin A in damaging human colonic epithelium *in vitro*. J Clin Invest 95: 2004-2011

25. Surawicz CM, Elmer GW, Speelman P, McFarland LV, Chinn J, van Belle G (1989) Prevention of antibiotic-associated diarrhea by *Saccharomyces boulardii*: A prospective study. Gastroenterology 96: 981-988

26. Toothaker RD, Elmer GW (1984) Prevention of Clindamycin-induced mortality in hamsters by *Saccharomyces boulardii*. Antimicrob Agents Chemother 26: 552-556

27. Triadafilopoulos G, Pothoulakis C, O'Brien M, LaMont JT (1987) Differential effects of *Clostridium difficile* toxins A and B on rabbit ileum. Gastroenterology 93: 273-279

28. Vidon N, Huchet B, Rambaud JC (1986) Influence de *Saccharomyces boulardii* sur la sécrétion jéjunale induite chez le rat par la toxine cholérique. Gastroenterol Clin Biol 10:13-16

29. Wilkins TD, Tucker KD (1989) *Clostridium difficile* toxin A (enterotoxin) uses Gal alpha-1-3Gal beta1-4GlcNAc as a functional receptor. Microecology and Therapy 19: 225-227

Conclusion

J. Thomas LaMont

Clostridium difficile. **Past, present and future**

Clostridium difficile is an organism with a curious past. The organism was discovered in the stools of healthy infants in 1935 by Hall and O'Toole who were studying the acquisition of the normal intestinal flora by newborn babies. The new organism was originally named *Bacillus difficilis* because it stubbornly resisted attempts at isolation. Later, the organism was placed in the genus *Clostridium*. Another curious finding in the initial report was the fact that *Bacillus difficilis* broth supernatant was highly toxigenic for laboratory animals. This was puzzling, in that babies harboring the organism were entirely healthy. Apparently, resistance of newborns to the organism explained their complete lack of symptoms. Shortly after its discovery the organism entered scientific "limbo" and did not resurface in the scientific and medical literature until the late 1970s.

The discovery by Bartlett's group that *Clostridium difficile* was the causative agent of antibiotic-associated colitis led to a remarkable resurrection of this pathogen. Rescued from obscurity, it burst upon the scientific stage as a new and very important pathogen. In the early 1980s, intense clinical laboratory investigations centered on the epidemiology, bacteriology, pathophysiology and clinical treatment of this interesting condition. Within a few years of its rediscovery effective treatment of acute antibiotic-associated colitis was provided as either metronidazole or vancomycin. In the mid-1980s attention was centered upon multiple reports of nosocomial outbreaks of *C. difficile* diarrhea. Hospitals in North American and Europe reported large outbreaks and several careful epidemiologic studies documented that upwards of 20% of hospitalized patients were infected with this organism.

A major breakthrough occurred in the first half of this decade when the genes encoding the toxins of *C. difficile* were cloned and sequenced. These studies showed that the two toxins of *C. difficile* are, in fact closely related with considerable amino acid and structural homology. The toxins are large protein molecules without subunits with molecular weights of 308 kD for toxin A and 275 kD for toxin B. Both molecules are potent cytotoxins and are lethal when injected into experimental animals, including primates. Although toxin A is more potent in animals in causing enterocolitis, recent evidence suggests that toxin B is more potent in humans.

The toxins are unusual, in that they do not resemble typical bacterial enterotoxins. The receptor binding portion of both toxins consists of tandem repeats at the carboxyl terminal one third of the molecule. Receptors for the toxins are not yet defined, although preliminary evidence suggest that intestinal brush border glycoproteins and erythrocyte glycolipids containing a terminal a-galactose are involved in binding. Recent investigations in newborn rabbits suggest that the toxin receptor is absent at birth and slowly develops after weaning. This might explain why human newborns are resist to *C. difficile* despite the fact the 50% of them harbor pathogenic organisms in their feces.

The most recent scientific breakthrough regarding the toxins relates to their cellular mechanism of action. Investigators in Germany, Paris and Boston have reported that toxins A and B possess glucosyltransferase activity directed at small GTP-binding proteins of the *rho* family. These proteins with average molecular weights of approximately 21-23 kD are critical mediators of actin filament formation in cells. The toxins inactivate *rho* A by adding a glucose residue to threonine in position 37°. This addition of a glucose inactivates *rho* and renders it incapable of interacting with actin. The end result is loss of actin filaments and rounding of the cell. It is not yet clear how this cellular mechanism causes diarrhea or colitis, but studies with cultured intestinal cells indicates that *rho* may be required for control of tight junctions which in turn regulate fluid and electrolyte secretion.

What does the future of *C. difficile* research look like? *C. difficile* is not likely to go away until we develop effective control measures, including vaccines. The organism is ubiquitous in hospitals and serious outbreaks are still quite common. In addition, endemic infection is widespread especially among elderly, frail patients. A number of unsolved questions regarding pathophysiology remain unanswered. For example, the involvement of the immune response in infection is still undefined. Serum antibodies directed against the toxin are present in more than 60% of healthy adults, but, it is not yet clear if these antibodies are protective. Coproantibodies are also present in intestinal secretions, but, their role in preventing or ameliorating infection is not yet known. Vaccination of animals with toxoids of toxin A or a combination of toxins A and B provides protection against subsequent challenge. However, no convincing studies are yet available in humans. Another serious consideration for the future is development of antibiotic resistance by *C. difficile*. At present, no antibiotic resistance strains have yet been described. However, a great concern among infectious disease consultants and bacteriologists is that *C. difficile* will become resistant to vancomycin or metronidazole. Antibiotic resistance could be disastrous, especially in debilitated patients with serious acute infection.

Another unsolved mystery is recurrence of *C. difficile*. A subset of patients, (approximately 10%) develop multiple recurrent infections whenever antibiotics are stopped. These patients may have relapses over many months or even years. A number of interesting treatment regimens have been proposed for such patients and several are described in this monograph.

I predict that the future of *C. difficile* will be marked by the availability of more effective treatment measures to control the disease. Passive and active immuniza tion are being tested by several groups. Clinicians will be better able to treat their patients who are suffering from acute infection. As an optimist, I feel that we will gain control of this infection by the turn of the century. Pessimists would predict that the organisms will keep the upper hand by acquiring antibiotic resistance genes from other enteric pathogens. Thus, the race is on between scientists and this very old organism to see who gets to the finish line first. Perhaps at our next meeting we will be able to write the final chapter for this unusual pathogen.

Achevé d'imprimer par Corlet, Imprimeur, S.A. - 14110 Condé-sur-Noireau (France)
N° d'Imprimeur : 20352 - Dépôt légal : novembre 1996 - *Imprimé en C.E.E.*